IMPORTANT NOTICE TO THE READER

During page makeup by the publisher, the following errors inadvertantly occurred. You should write these corrections in the book before beginning. This will greatly facilitate the use of the program.

Page Number

21	Turn to page 19
28	Turn to page 30
29	Turn to page 31
52	Line 4 should read: the matrix on page 53
	Last two lines should read: Compare your answers with ours on page 54.
53	Add: Turn to next page
54	First two lines should read: . . . in our Structuring Quiz on page 53
56	Add: Go to next page
58	Asterisked line should read: Our answers can be found on page 62.
69	Para. 2, lines 2 & 3 should read: . . . outlined on page 72.
	Para. 3, line 1 should read: . . . budget on page 72.
80	Line 5 should read: . . . our drawing on page 81.
87	Add: Turn to next page
96	Add: Go to next page
104	Add: Go to next page
110	Add: Go to next page
116	Add: Go to next page
118	Add: Go to next page
121	Add: Turn to next page
135	Para. 2, line 3 should read: . . . turn to page 139.

Charles E. Merrill Publishing Company

An Introduction to PPBS

JOSEPH H. McGIVNEY
Syracuse University

ROBERT E. HEDGES
Illinois State University

CHARLES E. MERRILL PUBLISHING COMPANY
A Bell & Howell Company
Columbus, Ohio

MERRILL'S SERIES FOR EDUCATIONAL ADMINISTRATION

Under the Editorship of

DR. LUVERN L. CUNNINGHAM, Dean

College of Education
The Ohio State University

and

DR. H. THOMAS JAMES, President

The Spencer Foundation
Chicago, Illinois

Published by
Charles E. Merrill Publishing Company
A Bell & Howell Company
Columbus, Ohio 43216

Copyright © 1972 by Bell & Howell Company. All rights reserved. No part of this book may be reproduced in any form, electronic or mechanical, including photocopy, recording, or any information storage or retrieval system, without permission in writing from the publisher.

International Standard Book Number: 0-675-09087-3

1 2 3 4 5 6 7 8 9 10—76 75 74 73 72

PRINTED IN THE UNITED STATES OF AMERICA

Contents

INTRODUCTION	1
SYSTEMS: A WAY OF THINKING	10
PLANNING	31
PROGRAMMING	48
BUDGETING	60
IMPLEMENTATION GUIDE	85
JARGON GUIDE	129
EVALUATION: SELF-TEST	135

Introduction

During the 1960s a crescendoing cacophony of critics decried both the rising level of governmental spending and the lack of explanation about the benefits to be derived from such spending. A partial response to this criticism was a 1965 White House Executive Order that directed federal agencies to adopt the systems approach to planning, programming, and budgeting. By 1967, President Lyndon B. Johnson's budget message to Congress included a special section devoted to the "Planning, Programming Budgeting System" (PPBS).

These developments at the federal level have heavily influenced and reflected the growing demand for accountability. Many state and local governments have adopted PPBS or program budgeting approaches in response to these demands. In California, Connecticut, Florida, New York, and Wisconsin, for example, the developmental aspects of PPBS are well under way.

Recently the effects of these adoptions have begun to extend to the local school systems. Thus it is desirable for everyone who has an interest in education to gain some understanding of the concepts and philosophy of the systems approach to planning and budgeting.

Most of the "systems approaches," whether they are called PPBS, Systems Analysis, Cost/Benefit Analysis, Program Budgeting, and so on, have as their major objective the improvement of the decision-making process through the application of critical thinking and scientific methods. Within this broad framework there are critical concepts and procedures that are applicable to most school systems.

For purposes of this book, the term Planning-Programming-Budgeting Systems (PPBS) will be used to represent the systems approach to planning and budgeting. Our aim is to convey a general feeling for the systems approach to planning and budgeting. The content of this book is introductory and includes material not always generally known to the neophyte. Thus, we will limit our discussion to some of the critical concepts fundamental to systems analysis and PPBS while providing the reader with opportunities to develop his understanding of these concepts and their relationships.

To convey a general feeling for the systems approach, we have attempted to combine several pedagogies including a variation on the programmed instruction format; a nonpedantic, informal tone of writing; a form of role playing; reinforcement for selection of "correct" answers, and benevolently intended sanctions for "incorrect" answers. Moreover, for those who want to try PPBS in their district we have included an implementation guide to serve as a first approximation of the events and processes leading to adoption. If thoughtfully studied the implementation guide could serve as an introduction to the preparation of flow charts.

We hope also to motivate the reader to further develop and refine his knowledge of PPBS and related concepts both in theory and in practice. Thus, the Jargon Guide includes not only definitions of concepts and terms used in this book, but also definitions of other terms frequently associated with PPBS.

One final note of caution. When the reader successfully completes this book, he will not be an "expert" in PPBS (unless he was before he started this book). There are many subtle, complex, theoretical, and practical aspects to which this introductory text will give little attention. For the reader wishing to learn more, we refer him to the sources listed on page 6.

Introduction 3

How To Use This Book

Role playing and person-to-person interchange of questions and answers have long been recognized by educators as valuable learning methods. In an attempt to capture the value of these teaching methods, the authors are using the technique of programmed instruction which has proved to be an excellent learning aid.

As you proceed through this book, you will be asked to assume several roles and be required to respond accordingly to the dialogue and questions presented by the instructors (authors). One of the first things you will notice is that this book is put together differently from a regular or traditional book. The pages are not read consecutively and, therefore, a bookmark will be helpful.

This format was chosen because the program was designed with variations in individual needs in mind and allows each reader to go through the material in a different order and in a different amount of time.

To proceed through the program, you need only follow directions. They are as follows:

1. Read the material on the page carefully.
2. If there is a question at the bottom of the page, turn to the page indicated directly after the answer you have selected.
3. If there isn't a question at the bottom of the page you will be directed to another page as follows:

Turn to next page

Like PPBS, this book is a change from the traditional approach. The questions and answers are designed to help you learn about PPBS rather than being a test. Don't be discouraged if your choice of answers occasionally leads you astray. We may have done it on purpose. Our goal is to have you break from traditional thinking and to make your decisions in a more rational manner.

Here is your first decision:

I would like to learn about the critical concepts and methods of PPBS from:

A traditional textbook. . . . turn to page 6

A programmed textbook. . . . turn to page 7

Introduction 5

THINK! .... You didn't follow instructions! Nowhere in this book will you find directions to turn to this page. What's more important, you turned to this page because you traditionally always turn to the next page in a book. Put the brakes on Traditional Thinking! Make the break from traditional thinking and take the modern approach ....

Turn to page 3 and read the instructions again

You have indicated that you would like to learn about the critical concepts and methods of PPBS from a traditional textbook. Let us suggest the following sources:

1. McGivney, Joseph H., and Nelson William. *PPBS for Educators: A Training Outline.* Columbus, Ohio: Center for Vocational-Technical Education, Ohio State University, 1969, 282 pp.
2. ―――. *PPBS for Educators: A Case Problem.* Columbus, Ohio: The Center for Vocational-Technical Education, Ohio State University, 1969, 170 pp.
3. ―――. *PPBS For Educators: An Annotated Bibliography.* Columbus, Ohio: Center for Vocational-Technical Education, Ohio State University, 1969, 60 pp.

This three-volume work was designed to help educators understand and train other educators in the essentials of the approach; it is available in hardcover volumes from the Center for Vocational and Technical Education, The Ohio State University, Columbus, Ohio, and from the E.R.I.C. system in microfiche.

We mentioned earlier that the major objective of PPBS was the improvement of the decision-making process through the use of critical thinking. If you look carefully at your decision not to go through this programmed textbook, you will see that your decision may not have been wise, particularly if you have paid for this book. Your decision not to use this programmed textbook should have been made before you purchased this book. Moreover, if you have already purchased this programmed textbook and have spent your money (resources) on it, the price is rather high for just three sources.

The point is that by the criteria that have been mentioned, your decision was not logical (although we recognize that your logic might have been sound if based on other criteria). Perhaps this is the way that you always purchase books. In this case, buying books and deciding to use other books first constitutes traditional thinking for you.

Turn to next page

Introduction 7

Your decision to break from tradition and use a programmed (this) textbook was a wise choice. We say this because we assume you bought this book to learn the critical concepts and methods of PPBS. To choose to read a traditional textbook on PPBS at this time would have been a less logical choice since you have started reading this book. Thus, according to the criteria stated above your choice was more rational.

Making rational decisions about allocating resources is what PPBS is all about. Let us offer the following definition of Planning-Programming-Budgeting system (PPBS):

> PPBS is a methodology for improving decisions that have to do with the allocation of scarce resources to attain maximum satisfaction of our unlimited wants.

More simply stated, PPBS aids in the attainment of efficiency. Let us examine the concept of efficiency as it relates to PPBS.

In our definition, we mentioned scarce resources. PPBS borrows from economics the assumption that all resources are scarce. Scarcity may be measured on a continuum in the sense that some resources are more scarce than others. In other words, you treat resources like there is never enough of them to do everything you want to do. Efficiency in PPBS, then, is trying to decide how to get the greatest return (output) from the investment (input) of your limited resources.

PPBS is basically a method to:

Hold down spending. . . . turn to page 9

"Get more bang for the buck" . . . . turn to next page

Good for you! PPBS is basically a method to "get more bang for the buck." The concept of efficiency in PPBS is to attain the most value or highest net benefit (bang) from the resources (buck) you invest. Any PPBS decision about the allocation of resources should focus on net value added or net benefit. Usually most people think of efficiency only as a means to hold down spending and thus they might be hanging on to traditional thinking.

To illustrate further, assume (1) you have limited funds with which to feed a large family for a week, (2) no matter how you spend the money they would never be well fed, and (3) your purchases of food could result in either starvation or subsistence. When you go shopping for food, you are trying to get the most value from the money (resources) available to you. By going to various markets, you could take advantage of the bargains at each market. The savings you attained on the bargains were thus available to be spent on additional food items since you were obliged to spend all your money on food. By being efficient (saving on the bargains) you were able to provide subsistence for your family rather than have them starve. Your efficiency was high because you got the most *value* or net *benefit* for your money—the family did not starve but subsisted. Thus, you did more than hold down spending; you made the optimum use of your resources because you got the best return from their use.

The efficiency concept in PPBS is the same. PPBS is basically a method designed to get more pay-off for the peso.

Turn to page 10

Introduction

You said that PPBS is basically a method to hold down spending. It is likely that most people agree with you and think of efficiency only as a means to hold down spending. We disagree and believe you may be hanging on to traditional thinking. Let us illustrate the efficiency concept by drawing on an example which most of us can identify with and understand.

Assume (1) you have limited funds with which to feed a large family for a week, (2) no matter how you spend the money they would never be well fed, and (3) your purchases of food could result in either starvation or subsistence. When you go shopping for food, you try to get the most value from the money (resources) available to you. By going to various markets, you could take advantage of the bargains at each market. The savings you attained on the bargains would be available to be spent on additional food items since you were obliged to spend all your money on food. By being efficient (saving on the bargains) you were able to provide subsistence for your family rather than have them starve. Your efficiency was high because you got the most *value* or net *benefit* for your money—the family did not starve but subsisted. Thus, you did more than hold down spending; you optimized the use of your resources because you got the best return from their use.

The efficiency concept in PPBS is the same. PPBS is basically a method to get more bang for the buck.

Turn to next page

Systems: A Way of Thinking

A Fable

Once upon a time a handsome boy and a pretty girl got married and lived in a rented house. They lived a happy life in their rented home with only one exception. They wanted to have their own "dream" house . . . a reasonable goal.

Hour upon hour the newlyweds would talk about their "dream" house and what it would be like to own it. Soon it became apparent they would have to act in order to attain their goal. They decided if they used their money wisely (as tradition would suggest), they could save enough money for a down payment in a year.

Together they decided on a plan and set out to accomplish their goal. Since she knew how to sew, they bought a sewing machine (on sale of course) to save money on clothes. A few months later, they purchased a television set at a bargain price to save money on entertainment. A short while later, they made a good deal on a freezer at a good price to save money on food. Toward the end of the year, they bought a washer and dryer at discount prices to save money on the laundry. When they went to buy their "dream" house at the end of the year, they found that they were broke. Their goal of achieving a down payment was not met.

The lessons of the story are:

1. Newlywed appliance needs should have been considered in their plan.
2. A goal gives us direction, but we should evaluate frequently to see to what extent our activities are helping us achieve our goal.
3. Getting more bang for the buck is not an end in itself.

As implied in our fable, the attainment of efficiency without directly relating it to goals does not necessarily improve our decision-making processes. There is a discipline, which has evolved mainly outside the field of education, that concerns itself with improving decision making. This discipline is known as the systems approach and PPBS employs

Systems: A Way of Thinking

many of its principles. The power of this discipline is that it offers a more solid foundation for integrating the many facets of decision making.

In its most simple form the systems approach consists of three interrelated notions: (1) input (the resources available to reach a goal), (2) process (the particular mix of the inputs), and (3) output (the product or the outcome of the process). Graphically this input-output system can be shown as follows:

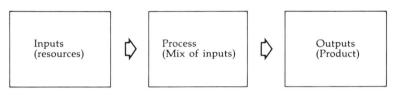

In such a system, we input resources, process the inputs in a certain way, and out comes a product or output. Do you believe this input-output approach can be applied to local school systems?

The input-output approach seems dehumanizing and thus cannot be applied to school systems . . . . turn to next page

The input-output approach may have great utility if applied to school systems . . . . turn to page 13

Sorry about that. Like it or not the input-output approach is being applied to many school systems. Many state legislatures and local school boards have mandated its use. The state legislature and local taxpayers "input" resources and out come students with varying amounts of education (output). In the process, the inputs (usually money which is used to pay for resources such as students, teachers, buildings, materials, etc.) are intermixed in various ways (teaching-learning processes) which collectively result in the outputs—the educated students who leave the system with varying types and levels of education.

Turn to page 14

Right you are!!! The monies of the state legislature, the federal government, and local taxpayers become the "input" of the system. The money input is transformed into other resource inputs such as teachers, materials, buildings, and so on. These inputs interact with students in the teaching-learning processes and result in the school systems' output —the educated students who leave the system with varying types and levels of education.

Turn to next page

Selected Features of the Systems Approach

The simple input-output approach we have just described may be of great use in itself in helping you make more rational decisions in the allocation of scarce resources. However, it could be expanded into a very complex approach. Since one of our goals in this book is to strive for clarity with simplicity and, since general systems theory permits flexibility in dealing with problems, we have selected some critical aspects of general systems theory for use in this book. We believe these features will be useful in helping you make your decisions in a more rational and systematic manner.

For our purposes, then, the systems approach includes the following features:

1. *Need*—Our newlyweds in the fable had a need.
2. *Goal stated as Objectives*—As our fable implied, a goal is (traditionally) a vague aim which sometimes provides us with nothing more than direction. The systems approach requires that the goal be redefined in precise and measurable terms which are referred to as objectives.
3. *Alternatives and Objectives*—Different ways for the newlyweds to obtain their goal. (They didn't have any alternatives identified.) The systems approach requires that alternatives be generated so that there is a choice.
4. *Alternative Methods*—The advantages and disadvantages of one set of alternative activities weighed against other alternatives or sets thereof. Like PPBS, the systems approach embraces our concept of efficiency. There will almost always be various methods or means for achieving an objective.
5. *Evaluation*—Unlike our newlywed's method, the systems approach requires that evaluation take place frequently.

Turn to next page

NEED

For our purposes, the authors define need as the difference between what exists and what is intended or required. The discrepancy between the present state or condition and what is intended is need, and quite often it can be measured. In the case of our newlyweds, the difference between their bank balance (we suggest that it was zero) at the time they implemented their plan and the amount they needed for the down payment was measurable.

However, all needs are not that easy to measure in such precise terms. Often we are able to judge only in gross terms whether or not there is a need. For instance, we explained the concept of efficiency as getting more bang for the buck earlier in the book. We also stated that PPBS is a useful approach in implementing and applying the efficiency concept to decisions involving the allocation of resources. With this information, you should be able to make the following judgment:

> In the face of "taxpayers' revolts," budgets getting larger, school problems getting more complex, increasing competition for government funds, and a growing number of critics of the educational system:

> *A need exists in school systems for PPBS. . . . turn to next page*

> *A need does not exist in school systems for PPBS. . . . turn to page 17*

Thanks for joining up with the authors. This is where it's at. In recent years evidence abounds that the availability of resources has not kept pace with those needed to solve the many problems confronting our nation. Resources to meet the problems of poverty, drugs, pollution, housing, transportation, national security, and education are inadequate. This emphasizes the need for local school systems to answer the question:

> How can we best use our resources (dollars) to attain optimum value added to the learning of students?

That is, how do we relate input to output to help students "learn" the most. Beyond that question is the need to demonstrate that additional resources, if and when they become available, would be more efficiently utilized by education than other worthy programs. In essence, there is a need for PPBS in school systems because educational decision makers can no longer rely exclusively on what amounts to seat-of-the-pants (traditional, if you prefer that term) methods.

Turn to page 18

Systems: A Way of Thinking 17

Whoops! We won't say you are wrong, but there may be several reasons why you turned to this page. The reason that we prefer is that you have based your decision upon your belief that your local school system is already making optimum use of its educational outputs. If this is so, we would like to visit this exemplary school system.

An exemplary school system like yours is envied by many. Its educational significance is great if its secrets can be learned and exported to other schools. Please contact us through the publisher.

But really folks, we hope you will overlook our mild but sarcastic sanction and stay with us.

On the other hand, however, perhaps you are not aware that the availability of resources has not kept pace with those needed to solve the problems facing our nation. Resources to meet the problems of poverty, drugs, pollution, housing, transportation, national security, and education are inadequate.

Therefore, local school systems should utilize resources as efficiently as possible. That is, they should relate input to output to help students "learn" the most. We believe that this can no longer be done by what amounts to seat-of-the-pants (traditional if you prefer that term) methods, and thus there is a need for PPBS in school systems. If this is the case we urge you to stay with us.

Turn to next page

GOALS AND OBJECTIVES

As our fable implied, a goal is traditionally a vague aim which sometimes provides us with nothing more than direction. After all, if the newlyweds had broken their goal into monthly bank balances (objectives), they might have had "signposts" or "milestones" to help guide them to their goal.

When a goal is redefined in precise and measurable terms, it becomes an objective. A good objective must meet certain criteria. It must be stated in qualitative and quantitative terms; it must include explicit information on how much of something will be done as well as how well it will be done. This is necessary so that we can measure and evaluate whether or not we are obtaining our objective.

If our goal was "to have a good science curriculum" then a good objective would be:

Some sixth-grade students should pass the X science exam at the end of the year.
... turn to page 21

All sixth-grade students should pass the X science exam at the end of the year. ...
turn to page 22

Systems: A Way of Thinking

ALTERNATIVE OBJECTIVES

The systems discipline requires that alternative objectives be generated. Without alternatives you are fostering a false sense of decision making because without choice there is no need for a decision maker. Usually, you will find that many potential outcomes (objectives) can be described and quantified. Often they can be ordered on a continuum ranging from least desirable to most desirable outcomes.

With only one outcome to consider, you'll never know if there is something else which will get you more sauce for the sawbuck.

Seeking alternative objectives is necessary and can prove to be enjoyable in practice. To become more proficient and to have some fun, try it in the social scene. For example, at your next party or group gathering, try the following responses to some of your friends' statements:

1. What are the alternatives?
2. Compared to what?
3. Yes, go on.
4. What else?
5. _____(Supply your own variations.)

Turn to next page

Alternative objectives are necessary in decision making. Generally, alternative objectives are compared with what presently exists. Often, however, decision makers lack a clear description and understanding of what actually exists. Even existing practices can be improved by better description. But beyond present practice, additional objectives should be developed to help understand and decide what can be. Often, a wide range of alternative "can be's" can be described, developed, and compared.

Therefore:

Alternative objectives provide the decision maker with a choice. . . .
turn to page 23

At least five alternative objectives for each decision must be developed. . . .
turn to page 24

We can't quite agree with you. You chose "*some* sixth-grade students should pass the X science exam at the end of the year." *Pass* is the qualitative portion of both statements. *Some* and *all* are the quantitative portions of the objectives. *All* is more precise (and thus more measurable) than *some*. Our preference would be "*All* sixth-grade students should *pass* the X science exam at the end of the year."

Turn to page 22

Good Show! *All* is the quantity and *pass* is the quality in that objective. The objective is good because it is explicit both quantitatively and qualitatively. Thus, progress toward the objective can be measured.

It should be noted that objectives can also be thought of as the benefits that you wish to obtain on the way to your goal.

Turn to page 19

You chose "alternative objectives provide the decision maker with a choice." You are right! This is one of the functions of alternative objectives.

Turn to page 25

You chose "at least five alternative objectives for each decision must be developed." We are sorry, but you are wrong. Although there is no maximum limit to the number of objectives that can be developed, at least one alternative is necessary. By having at least one alternative objective the decision maker is provided with a choice. This is one of the functions of alternative objectives and PPBS.

Turn to next page

Alternative Methods

The systems approach also requires that you develop alternative methods (or different mixes of resources) to achieve each objective. Eventually the decision maker must decide which method or sets thereof will give him the most net value over the entire life of the proposed program. Accordingly, in addition to direct outlays of money he must also consider interest or the cost of money.

One plan of attack is quite simple. Cost out each method to find the least costly method of obtaining a given objective. Ideally, the advantages or disadvantages of alternative methods are measured in monetary terms and the cost of money (interest) is also considered. Below we have constructed a problem of sum interest.

If Method A would cost $25,000 annually for four years and Method B would require $60,000 during the first year, $30,000 the second year, and $5,000 in each of the third and fourth years to achieve the same objective, then:

Method A is the best choice. . . . turn to next page

Method B is the best choice. . . . turn to page 27

You are right! Spend as little as possible for the same or equivalent results. We are trying to get more force for the Franc. In arriving at this choice, you probably considered the fact that resources have alternative uses. Since money is a resource it also has alternative uses. Usually the cost of money is measured by the interest one has to pay for its use. If you computed the cost of interest at any realistic rate (1 percent to 10 percent), you would have to decide that Method B cost more than Method A to achieve the same results in the same four-year period. The difference in first-year outlays alone ($35,000) computed at 1 percent would save $350 in the first year. Similarly, the difference ($40,000) in the sums of the first two years ($60,000 + $30,000) versus ($25,000 + $25,000) would result in additional savings of $400 assuming a 1 percent interest rate.

By now you've got the idea.

Turn to page 28

Perhaps you chose Method B because the authors did not adequately convey to you the notion that money is a resource and thus has alternative uses. If so let us try to clarify this misunderstanding. Usually the cost of money is measured by the interest one has to pay for its use. If you would have considered the cost of interest in making your decision, you most likely would have selected Method A in preference to Method B. With any interest rate (above zero) Method B would cost more than Method A to achieve the *same* objective. Thus, Method A would be the best choice. A more detailed explanation of the cost of money (interest) can be found on the previous page.

Turn to page 26

EVALUATION

The systems discipline is not a set, established system which when once set in motion unerringly seeks its objectives; rather, it is a dynamic process in which virtually all sets of relationships are subject to change. To insure that objectives are still valid and that the methods for achieving them are producing the expected results, evaluation is necessary.

Evaluation requires frequent checking. As for us, let's now evaluate. To see if you are following us, try the quiz below.

MATCH THE FOLLOWING

    _____1. Need   (a) punch for the penny

    _____2. Efficiency   (b) gives us choices

    _____3. Goal   (c) broad statement providing direction

    _____4. Objective   (d) choosing the best method to achieve an objective

    _____5. Trade offs   (e) aims stated in qualitative and quantitative terms

    _____6. Evaluation   (f) difference between what is and what should be

    _____7. Alternatives   (g) frequently checking our progress toward our objectives and goals

Turn to next page 30

Systems: A Way of Thinking

SYSTEMS AND PPBS

What you have just learned about systems can be utilized in developing your knowledge and skill in PPBS. In brief, the major contribution of the systems discipline to PPBS is its potential for integrating the processes of planning, programming, and budgeting.

Planning starts with identifying need—the gap between what is and what should be. It includes the setting of goals, determining objectives, and developing alternatives to these goals and objectives.

Programming is basically a process for examining potential trade offs. It requires that we develop and analyze alternative methods to achieve each objective including our alternative objectives. Accordingly, we try to look at several different combinations of inputs (money, teachers, students, buildings, equipment, etc.), each of which will help us reach our objective(s). We then try to make valid comparisons between and among the several different methods and pick the best one.

The budgeting process is the legal and financial means for allocating resources. Budgets are usually authorized for one- or two-year periods. They set forth the total and specific estimates of what resources may be used by a given agency such as a school board. In PPBS, we view the budget as the official approval for the first year of a multiyear program. Each year we utilize our experience to adjust and change our multiyear programs.

Evaluation, of course, pervades each of the above steps. To set goals the gap between what is and what is desired must be identified, that is, evaluation. To optimize the mix of inputs also requires evaluation of alternative methods. Evaluation of the effectiveness of actual expenditures in terms of expected accomplishment usually takes place after budgets have been executed.

Turn to next page

Answers to Matching Quiz

1. Need   (f) difference between what is and what should be
2. Efficiency   (a) punch for the penny
3. Goal   (c) broad statement providing direction
4. Objective   (e) aims stated in qualitative and quantitative terms
5. Trade-Offs   (d) choosing the best method to achieve an objective
6. Evaluation   (g) frequently checking our progress toward our objectives and goals
7. Alternatives   (b) gives us choices

Turn to page 29

Planning

Earlier in the book, we promised you a chance to do some role playing. Well, now is the time. Because of the limitations of this medium, we will have to approach it in a slightly different manner from that which tradition usually demands.

In the following pages, the authors have written scenarios from their combined experiences to give you a chance to practice your newly found knowledge. At the end of each scenario, we will give you a decision to make. We want you to make your decision by drawing upon what has been presented so far and from the perspective of the role that we will indicate for you at the beginning of each scenario.

Instead of getting up before a group and playing your role, you should play it in sort of a "Walter Mitty-ish" way, that is, privately. Be a good "method" actor and throw yourself into the part. Be empathetic with the "character" that has to make the decision! Draw upon experiences, emotions, and so forth, to play the role.

Our strategy is to have you make as many decisions as possible from other persons' points of vantage. We think that you will gain some insights into the application of what we have been advocating from perspectives different from your own. Hence, we have endeavored to mix it up just to be sure.

Turn to next page

Acknowledgments

Tradition generally dictates that this item be placed in the front of a book. But, as you may have surmised, we are generally not governed by tradition if it doesn't help us attain our goals. We have several goals that need to be attained at this point in the program.

The authors would like to thank all those people who have helped make the scenarios possible. Without them this book could not have been written.

Since we have been informed by some of our associates, who read earlier drafts of this book, that the scenarios may reveal that certain types of schoolmen and schoolwomen possess certain characteristics which Silberman classified as mindlessness, we must point out that this is not intended in this book. Indeed, we hereby emphasize our belief that each scenario pertains to a unique incident and individual. In no event should a scenario be generalized to classes of roles or groups.

In addition, the authors have changed names, places, and so on, to protect individuals and groups who are innocent. To also be rational, we have done it to protect ourselves.

Moreover, our gratitude extends to our students, colleagues, and friends whose careful reading and thoughtful suggestions have improved the quality of this undertaking.

Finally, we dedicate this book to our spouses, Margaret Quandt McGivney and Carol Foster Hedges, whose most important contribution of spiritual and emotional sustenance, while not completely measurable, was timely and seemingly boundless.

Turn to next page

Planning

SCENARIO 1

Your role: Parent

Background: In School District X, there has been a steady decline in the number of high school students who have been selecting a foreign language as part of their high school program. In an effort to offset this trend, school officials proposed to offer a foreign language program in the elementary schools. This caused some controversy in the district and the parents divided into several camps over the issue, including a group that was concerned with the proposed increased expenditure for the program. The foreign language director of School District X was invited to speak to all interested groups. The following occurred:

> *Foreign Language Director.* . . . As you can see, most of my remarks so far have been directed toward the need of students in high school to take a foreign language. But your primary interest is in the proposed foreign language program in the elementary school. Foreign language instruction in the elementary schools has gained wide acceptance during the past two decades. The objective of such a program is as follows:
>
>> That students in the elementary school should develop a skill in reading and writing and the ability to communicate with others in a foreign language.
>
> *This is a goal.* . . . *turn to next page*
>
> *This is an objective.* . . . *turn to page 35*

Welcome. You decided it is a goal. You're right. The foreign language director may have called it an objective, but it is not an objective in PPBS. In order for it to be an objective it must be stated in measurable terms.

Another way to think of an objective is to have it represent the expected outputs of a system. In our earlier model of a school system we noted that resources were input and the outputs were students with varying amounts of "value added" in education. In other words, the processed students were our expected outputs.

Turn to page 36

Really? That collection of words is an objective? Perhaps you were misled by the fact that the foreign language director used the word objective. However, our criterion for an objective in PPBS is that it must be stated in measurable terms. In our opinion, the foreign language director was really stating a goal or vague aim of the program.

Think of the program as a model system. Money (resources) is going to be input and the outputs are going to be students with varying amounts of "value added" in foreign language skill and knowledge. In other words, the processed students are the expected outputs of this program. Therefore, an objective of this program should be stated in measurable terms so that we know if the program is giving us the expected output.

Turn to next page

Scenario 2

Your role: Teacher

Background: In School District W, it has been the policy of the school to test preschool children before they enter kindergarten to determine if they have matured enough for school. If the child scored below a certain level on a maturity test, the parents were advised to keep the child home for a year. However, most parents have not heeded the school officials' advice and have sent their children to school anyway. The school officials were obliged by law to accept the children. Normally, about thirty children were found to be immature. The school officials have now decided to set up a special class known as the Pre-Kindergarten Program to help relieve this problem. They are going to give the students a specialized program to help prepare them for kindergarten. This will require that the students receive as much individualized attention as possible. Therefore, the class size must be limited.

The objective of this program is to give fifteen immature students an individualized program to help them become ready for kindergarten.

This a goal. . . . turn to page 38

This is an objective. . . . turn to page 39

Planning 37

SCENARIO 3

Your role: School Superintendent, Mr. S—.

Background: In School District A, the results of standardized tests indicate that of the 1,200 students leaving sixth grade 50 percent of them are below fifth-grade level in reading. As superintendent of schools, you have hired a reading consultant to look into the problem and make some recommendations to you as to what might be done.

Reading Consultant:
>   Mr. S—, I think that the solution to the reading problem is to start a remedial reading program in the elementary schools. I've been visiting other school districts that have such programs and I believe that it would be an ideal solution to the problem we have here.

Mr. S—:
>   What were the programs like in the other schools?

Reading Consultant:
>   They were all basically the same. A remedial reading teacher in each elementary school worked with students on a one-to-one basis; the students were referred by their regular teachers.

Mr. S—:
>   Since we have ten schools, this means that we'll have to employ ten new teachers. What kind of results do you think this remedial reading program will achieve in the next few years?

Reading Consultant:
>   When the program becomes operational within two years, less than 10 percent of the students leaving the sixth grade each year will be reading at or below the fifth-grade level.

This is a goal. . . . turn to page 40

This is an objective. . . . turn to page 41

You decided it was a goal. You've got the idea. The school official gave us a little more information than the foreign language director about the quantity of students processed, but he really didn't say what the expected quality of the students would be at the end of the program. In PPBS, an objective must be stated in qualitative and quantitative terms.

Turn to page 37

Planning

You decided it was an objective. Sorry, but you're not correct. We've got to admit that we received a little more information than we did from the foreign language director. We do know the number of students that is going to be processed in the program, but we don't have an estimate of their expected quality at the end of their program. In other words, an objective to be in measurable terms, must be stated in qualitative and quantitative terms.

Turn to page 37

How did you get here? We tried to make it very clear what the difference between a goal and an objective was. The reading consultant gave an objective. The expected outcome or output of his program was that less than 10 percent (quantity) of the students leaving sixth grade would be reading at or below the fifth-grade (quality) level. We can certainly measure both of these. We assume the number of students can be computed (e.g., 1,200 × 10 percent = 120) and that some kind of reading test will be administered to determine grade level. Issue can be taken as to what test will best suit our purposes, but certainly not with the fact that such tests do exist, and that they have been used as measures of the output of the educational process.

Turn to next page

We agree. It is an objective. It meets the requirement of being stated in qualitative and quantitative terminology. Less than 10 percent of the sixth-grade class is the quantitative aspect and the qualitative aspect is the fifth-grade level in reading.

Turn to next page

SHORT SCENARIO

Your role: School Superintendent, Mr. S—.

Background: It's the same as Scenario 3. However, we will give you the added information that you have the money and the authorization to spend it on reading. As superintendent of schools, would you accept the reading consultant's recommendation?

Yes. . . . turn to page 45

No. . . . turn to page 44

Scenario Summary

We hope that you now recognize that the scenarios are not absurd. We believe that what we have described is fairly common practice in educational organizations. We hope that the scenarios have demonstrated how limited and narrow the decision-making processes regarding the allocation of educational resources can (and perhaps have) become. The scenarios demonstrate that the generation of alternatives may help widen the perspective of all those involved in decisions relating to the allocation of resources. For our purposes, then, planning requires us to develop and state alternative objectives in qualitative and quantitative terms.

As you may have guessed from reading the answers to the questions posed in the Short Scenario, the planning aspects of PPBS are not undertaken in isolation from its programming (and budgeting) aspects. That is, mere statements of objectives in quantitative and qualitative terms are not sufficient in themselves to provide a decision maker with all the information he needs to make the decision that is the most efficient. Rather, alternative sets of activities, with estimates of the cost of inputs and expected outputs, for each objective are extremely important in helping him arrive at efficient decisions. Conversely, alternative sets of activities unrelated to specific, explicit, and measurable objectives are also an insufficient basis for deciding upon the allocation of resources.

Remember we need alternative objectives but we also should explore alternative methods or activities for achieving the objectives. By doing both we are well on our way to getting more bang for the buck.

Turn to page 46

You're on the right track. You asked the reading consultant for recommendations; he offered a single solution. Send him back to do some more homework. A major tenet of PPBS is that consideration of alternatives improves the probability of making a "good" decision. Thus, he might find an alternative objective which might give approximately the same result (reduce the percentage of sixth graders reading below fifth-grade level from 50 to 15 or to 5). Perhaps of equal value is the prospect that generating alternative objectives will help generate alternative activities—other means and methods of achieving the same or equivalent objectives. Other sets of activities after careful analysis could turn out to be even more efficient. For example, if your objective was to reduce the 50 percent to 1 percent you may need fifteen additional teachers. Since you have support only for ten teachers you may want to examine alternative activities such as differentiated staffing utilizing subprofessional aides (whose salaries are much lower than professional teachers) supported by a heavier reliance on programmed instructional materials. We'll have more to say about alternative methods in the next unit on programming. But, for now, remember that alternative objectives give us a choice.

Turn to page 43

Planning 45

You would? Well we disagree. As superintendent, you asked the consultant to make recommendations. In other words, you asked for alternative solutions to your problem. He gave you only one and, from your response, he did a good selling job. The point is that you have accepted a single program which you expect will greatly reduce the percentage of sixth-grade readers achieving below grade level. Other approaches may have permitted you to reduce the level from 50 to 10 percent at a much lower cost. Moreover, the development of alternative objectives such as reducing the percentage of low achievers from 50 to 25 or even 5 percent may have stimulated the reading consultant to consider alternative means of reaching the objectives.

Consider for a moment that hiring ten teachers will cost about $70,000 annually (10 x $7,000 salary). There is growing evidence that subprofessionals, when provided with proper direction and proper materials, can efficiently supplement and complement the professional teacher—freeing him to devote more of his time to planning, organizing, coordinating, and directing the activities of the instructional team. Furthermore, since the salaries of paraprofessionals are considerably less than those of professionals, it may be possible to achieve a reduction of the figure 50 percent to 5 percent for the same expenditure or less.

Remember that alternative objectives give us a choice.

Turn to page 43

Break Time

After we have spoken of bang for the buck, etc., we should practice what we preach. We believe (contrary to the approach successfully employed by Professor Higgins on Miss Doolittle in *Pygmalion*) that your efficiency in learning about PPBS may drop off without a respite for repose. Based on our instructional experience with students at the university, in-service workshops, colloquia, and so forth, the authors recommend that you "take five" before we continue our lesson.

If you're sitting, stand up and walk around. If you're lying down, stand up and stretch. Have a beverage of your choice.

In any case meet us on the next page in five minutes where we'll elaborate the programming process.

Turn to next page

Planning 47

Dear Partner-in-Learning:

Welcome back from the break. We realize that prescribing a break in the middle of a book is not traditional. But, by now, we think that you should have an idea of how we view tradition. While our own tradition so far in this book would dictate that we would have you choose an answer and then turn to the page indicated throughout the entire book, we haven't done this in the chapter on programming. You see, even our own traditions are subject to change.

In considering the alternatives available in presenting the critical concepts of programming, we have chosen a more straightforward method —usually associated with the traditional style of professors.

Using Tradition to Break Tradition,
J. H. M. and *R. E. H.*

Turn to next page

Programming

Programming is the link between planning and the decision to authorize resources to implement programs. It is the exploration of alternative ways of achieving an objective (or to changing the objective itself). By drawing on your experience, you probably can recall situations in which you've thought of several different ways (means) to achieve an objective. Programming is much the same. We develop and analyze several different ways (means) to achieve an objective. And in the process we may find that we want to change the objective.

Does it sound simple? Well it really isn't! If programming is to be a functional link between planning and budgeting it requires the utilization and integration of knowledge and skill not usually resident in any one person. Accountants and budgeteers, computerites and curriculeers, demographers and demagogues, psychologists and statisticians, parents and publics, students and system analysts—all possess knowledge, skill, and insights of great value to the process of programming. Although we will demonstrate how some of these persons might be involved in implementing PPBS (see the chapter on implementation), to describe and explain the relationships of these skills to programming is far beyond the scope of this book.

Instead, we've chosen to describe and explain the importance and centrality of the program structure to the programming process. As we hope to demonstrate, the program structure not only facilitates the goals of programming but also provides a systematic framework for linking the planning and budgeting processes.

The goals of programming can be stated quite simply:

1. To put together different mixes of inputs (resources) for given objectives.
2. To compare various mixes of inputs for given objectives.
3. To aid the decision maker in selecting the most efficient mix of inputs for given objectives.

Programming

A program structure provides us with a framework that aids us in reaching these goals. And let's not forget that our initial efforts at planning help us in developing and refining our program structure. Remember that planning has supplied us with specific, explicit, and measurable objectives.

STRUCTURING—PHASE ONE

We feel that structuring is composed of six interrelated elements: Who-What-Whom-When-Where-How. The interrelatedness of each element is visually shown in our drawing, titled *Structuring*.

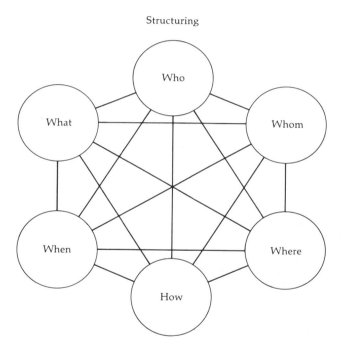

Structuring

Now let's go on to see how we can apply structuring to the more practical problems of educational endeavors.

In practice you'll have to supply your own meanings to these elements. As you move from the more general or abstract elements of who and what you must decide what "who" and "what" means to you. In health, "who" could be defined to mean patients; "what" could be defined to mean therapies or treatments. For our purposes let's (1) define our domain as education, (2) translate the six abstract elements to give them more precision in the educational domain, and (3) list four descriptors or examples of how each educational element could be further defined. We suggest you carefully examine the following Conversion Table to see how we have tried to give more precise meanings to the six elements.

CONVERSION TABLE FOR CONVERTING SIX ELEMENTS FROM ABSTRACT TO SPECIFIC MEANINGS

Six Abstract Elements	Education Meaning of Six Elements	More Detailed Descriptors of Meanings in the Educational Domain
Who	Student	Grade level, sex, age, mental health, etc.
What	Curriculum	Reading, music, physical education, art, etc.
Whom	Personnel	Teacher, psychologist, specialist, consultant, etc.
When	Time	Period, day, semester, year, etc.
Where	Location	Classroom, building, museum, home
How	Pedagogy	Lecture, lab, programmed text, CAI

From the Conversion Table you can see that we have attempted to keep our six elements, their educational meanings, and their more detailed descriptors mutually exclusive. That is one of the functions of a framework or structure; it helps us to sort out and classify our concepts and data in a systematic way.

But programming requires us to develop and analyze alternative mixes of inputs for each objective. Let us see how we may utilize the framework to facilitate our programming efforts.

Suppose we ask ourselves the following questions:

>*Who?* is going to get
>*What?* supplied by
>*Whom?*
>*When?*
>*Where?* and
>*How?*

As you will see, our framework provides us with a structure for answering these questions. But very often our programming task is made easier if our planning process has been adequately done. That is, planning supplies us with objectives. Most frequently objectives provide us with detailed descriptions of who and what. Unfortunately,

Programming

they frequently *do not* provide us with descriptions of whom, when, where, and how. To illustrate, let's reconsider an objective we discussed in the scenarios.

"At least 80 percent of the students completing sixth grade will be reading at the sixth-grade level."

This objective (like most others) provides us with answers to the *who* and *what* elements in our structure. In this case "who" is every student eligible to enter the sixth grade. "What" is the reading component of the curriculum. Thus, the planning process has greatly aided the programming process. By keeping our six elements in mind we now have specified the content of two of them?

Who .... students .... sixth grade
What .... curriculum .... reading component

This leaves us with the task of specifying and then combining the descriptors for the whom, when, where, and how elements. If this may appear to be quite simple, let's quickly agree that it is not. To illustrate the mathematical complexity we'll assume that each element is limited to the four descriptors noted in the Conversion Table. By utilizing all the potential combinations of options within and among each of the four descriptors for each of the four elements, theoretically there would be over 35,000 possible conbinations.

Don't despair! Structuring is not a substitute for the skills of highly trained personnel. The knowledge, wisdom, experience, and judgment of highly trained personnel can reduce the number of possible mixes to a workable number of viable alternatives.

For example, experience and knowledge may quickly tell us that, as soon as we specify "whom" (let's say regular elementary teachers), we greatly reduce the number of practical options available under the "when," "where," and "how" elements. Since a regular elementary teacher has other assignments (e.g., math-science) his "when" (periods, days, weeks) available for reading are limited. Similarly, "where" becomes limited because of the inefficiency of his traveling between buildings. In the same way "how" is limited by the pedagogical competencies of a given teacher. If we start with one of the other elements, those remaining similarly become limited. Practically speaking, then, knowledge, wisdom, experience, and judgment can simplify our programming difficulties.

Nevertheless, we believe that by rotating our elements and starting with a different one each time, a wider range of mixes (alternatives) will be generated. Thus, the structuring or programming process can be an heuristic device for developing alternatives.

Remember that the planning and programming processes are interdependent. Also keep in mind that objectives may be stated in quantitative terms that include elements other than "who" and/or "what." If so, your use of our structuring guide should be accordingly adjusted.

Given our rather general treatment of structuring we've developed the following matrix to summarize this phase of our structuring process. On the horizontal axis we've listed our six abstract elements. On the vertical axis we've listed our operational definitions of the elements in educational terms. In so doing we have created a matrix containing thirty-six cells. Only three of the cells have been filled in; three contain questions marks (?). See if you can fill in the appropriate descriptors in the cells with question marks (?). You should be able to get the answers from our previous discussion. Compare your answers with ours on the next page.

Programming

STRUCTURING QUIZ

	Who	What	Whom	When	Where	How
Students	?					
Curriculum		?				
Personnel			?			
Time				First two hours daily Monday–Friday, September–June		
Location					Cluster F in Horace Mann School	
Pedagogy						Lecture discussion

ANSWERS TO STRUCTURING QUIZ ON PREVIOUS PAGE

	Who	What	Whom
Students	Sixth grade		
Curriculum		Reading component	
Personnel			Regular elementary teachers

With the answers above added to those in our Structuring Matrix on page 49, we have shown one set of six descriptors combined to summarize one mix for one objective. We repeat once again that several alternative mixes should be developed within the structuring process. By referring to the descriptors in our Conversion Table (or by using your own) you should be able to develop alternative mixes for the given objective.

Costing the Mixes

For each mix we would then begin to estimate the monetary value (or dollar cost) while also keeping track of the "real" resources such as teachers, classrooms, and equipment. For example we might cost account 20 percent of a teacher's salary to one particular mix but we would have to remember that 80 percent of the "real" resource is left and should efficiently be used in some other mix. By employing acceptable prorating techniques and many of the concepts from economics it is possible to arrive at total costs for each of the mixes. But to do that in this introductory book is beyond the scope of the book. We ask you to accept this statement based on your own knowledge or accept it on faith.

Now let's get back to structuring.

Turn to next page

Structuring—Phase Two

So far, we have been developing a structure to facilitate the development and comparison of alternative mixes of resources to achieve one objective (e.g., sixth-grade reading). However, in PPBS, objectives are rarely considered in isolation from other objectives and programs. They should be related to the total educational program in the district. For example, sixth-grade reading can be considered to be but one segment of the total sixth-grade program; and it can be considered to be only one part of the total reading program for the school district. How do we deal with these interrelated "programs"?

Within any district, one could develop a hierarchy of levels and combine these levels into one total program structure. This can be done if we develop and assign our objectives to appropriate levels in the hierarchy or program structure. Therefore, if our objectives are to be dealt with in a systematic way it is necessary in programming to develop different levels for a total program structure.

As you may have anticipated, our framework or structure is useful in dealing with different levels of a program structure. Most school districts have explicitly or implicitly organized their program structures in terms of the six elements (students, curriculum, personnel, location, time, or pedagogy) or combinations thereof. Why? Because in any program structure for education, virtually all of these elements must be considered at some level. Usually, emphasis is given to one or two of the elements at each level in the structure. Moreover, it frequently happens that the emphasis placed on certain elements at one level will influence the elements stressed at another level.

To illustrate this, let us consider a hypothetical school district. At the school district level (that of the board of education and superintendent), it is often difficult to determine which elements are being stressed because the needs of the district are usually stated in very global terms. However, at the next level below the board of education and superintendent, the school district generally gives emphasis to one or two of the elements. The emphasis at this level will in turn influence the emphasis at the next level below it, and so on. The actual number of levels needed in any particular school district will be largely determined by its size such as its number of students, buildings, staff, and so on.

To provide you with a visual illustration of a program structure involving several levels, we've developed the program structure on the next page which portrays one way of integrating various levels within a program structure of a school district. This program structure is generally referred to as a grade-level structure because the relationships between levels of the instructional program are basically grade based. REMEMBER THAT THERE ARE MANY OTHER REALISTIC, VIABLE ALTERNATIVES TO THE STRUCTURE SHOWN ON PAGE 56!

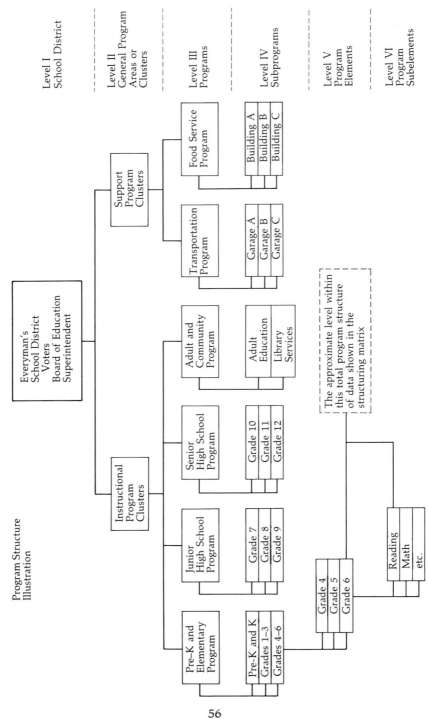

Programming 57

The chart on the previous page is an abridged version of a total program structure emphasizing "who" (student by grade level). Because of space limitations, we did not include several programs (level III) such as special education and summer school under the heading of Instructional Programs Cluster, and debt management and plant under the heading of Support Programs Cluster. In practice, all programs and subprograms should be included in the total Program Structure.

The authors believe that the development of such a structure is necessary to deal with the mind-boggling complexity otherwise associated with developing and comparing alternative sets of objectives and mixes. Even with an excellent program structure, the remaining complexity is great. Recall for a moment our earlier discussion of the potential alternative mixes for a single objective. This is further confounded by the interrelatedness of objectives at the different levels of program structure.

A program structure helps us analyze objectives and their mixes from a consistent perspective and provides a framework for costing out our mixes. In comparing alternative mixes for a given objective, we must also consider their utility to all objectives in the total program structure. Thus, while we can cost account 20 percent of a teacher to a given objective we must also remember that the remaining 80 percent should be efficiently utilized for other objectives. By utilizing the program structure as a dual memory bank for actual resources (a whole teacher) and their monetary equivalent (prorated costs) we can facilitate our goal of comparing alternative mixes.

The program structure also helps the decision maker in trading off alternative objectives. By establishing reasonably comparable levels within the program structure, he can compare the expected benefits and costs of one objective to several others with greater knowledge of the interdependence of the resource mixes supporting the objectives. As he starts making trade offs among objectives and resource packages he is more fully informed of the consequences of such trade offs.

Thus in programming, through structuring, we are able to (1) develop comparable levels of objectives within the structure, (2) put together different mixes of inputs for objectives, (3) compare various mixes of resources for given objectives, and (4) aid the decision maker in selecting the most efficient mix of inputs for given objectives.

As planning links us to programming, programming links us to budgeting. In the next chapter we will show how programming and the program structure greatly aid us in budgeting.

But before beginning budgeting, it's time to test the transformation in knowledge, attributable to our traditional teaching of programming.

Turn to next page

Programming Quiz*

A. Here's an evaluation instrument, called "the one-word fill-in." Most pedagogues should be familiar with it.

Drawing on your knowledge of structuring as presented in this chapter fill in the missing word for each of the six items listed below. (To help you get started we'll give you the first one.)

Educational Element	Abstract Element
1. Student	Who
2. Curriculum	_____
3. Personnel	_____
4. Time	_____
5. Location	_____
6. Pedagogy	_____

B. This is called multiple choice (or guessing). We'll present you with a statement and ask you to select one of the choices or answers listed below the statement.

IN ALMOST ANY PROGRAM STRUCTURE "SEVENTH-GRADE READING" WOULD BE PLACED AT A HIGHER LEVEL THAN WOULD "SEVENTH-GRADE READING AT THE HORACE MANN SCHOOL."

Yes____   No____   Maybe____

*Our answers can be found on the next page.

Turn to page 62

Budgeting

We now have at least a general notion of how planning leads to programming and how the latter often leads us back to planning. But programming also leads us to the budgeting process. In budgeting we actually decide to implement the program plan we developed in the planning and programming processes.

Whether or not budgeting is viewed as part of PPBS, it is *the* legal and financial means for allocating resources. The budget process actually authorizes the use of the resources by one or more public or private agencies. Historically most budgets have authorized the purchase of or use of resources without clear, explicit estimates of the expected outcomes.

Let's ask the question "What should a budget do and what are its functions?" The authors have already expressed the view that an educational budget should have the same end in view as the planning and programming processes. That is, the budgeting decision should be based upon the objectives of the organization and draw upon the wealth of data considered in the planning and programming processes. But the budget also has some traditional (and necessary) functions in its own right.

First, a budget should define the availability of monies to an organization. This is usually called revenue or income estimates. Your school district has a budget in which it estimates the amount of money available to it for spending for a given time period (usually one year). The second function of a budget is to specify what objects or services may be purchased within the given time period. This is usually called expenditure estimates. This is equivalent to specifying the types and kinds of people and things for which expenditures can be made. (Earlier we called this function input.) In PPBS, the budget takes on the additional function of relating these inputs to the objectives and outputs that have been described and explained in the planning and programming processes.

These thoughts may sound like common sense, and in fact they are to the extent that the budgeting process approaches the full potential of a PPB system (e.g., the requirements noted above). Yet, many budgets do not meet these requirements.

In fact, let's speculate how some school budgets are made up.

Turn to next page

Budgeting 61

BUDGET METHOD 1

As we noted earlier in our book, many persons confuse the concept of efficiency with the concept of economy. Some school boards do the same. Most of us have probably heard or read about school boards and superintendents that highly value the goal of economy. Let's eavesdrop on one such group as they go about making their budget.

> Establish spending limits! Clearly establish spending limits. It's very important that we keep spending in line with the past year. Since we're not sure if state aid will go up, stay the same, or go down we'd better estimate state aid conservatively. Federal aid levels have not been too reliable so we'd better be conservative in estimating the federal share. With high levels of inflation and unemployment, we probably can't expect the local taxpayer to support an increase in the real estate tax. If we increase his taxes he might vote our budget down. So let's not increase the local tax rate. However, once we establish the spending limits, we'll divide up the total and establish spending limits for each of the various departments. We'll just have to say to each department head:
>
>> These are all the resources you get, but you can use them as you see fit. This is what you have to work with this year. We expect you to use these resources to get the best possible return in terms of value added to the education of our children.
>
> *This method will produce a programmed budget. . . . turn to page 63*
>
> *This method will not produce a programmed budget. . . . turn to page 64*

Answers to Programming Quiz

Question A.

1.	Student	Who
2.	Curriculum	What
3.	Personnel	Whom
4.	Time	When
5.	Location	Where
6.	Pedagogy	How

Question B.

We think we understand how those of you who answered yes or no to this question could have "built your case" and thereby logically defended your decision. For example, if you *assume* that the instructional program cluster (see Level II on page 56) has been given a who (seventh grade)–what (reading) emphasis then it would follow that "Horace Mann School" (where) could be a logical level below that of "seventh-grade reading." Thus, a "yes" would be right. If, however, you *assume* that where (several attendance areas including "Horace Mann School") has been emphasized at Level II, then "seventh-grade reading" would logically be placed at a lower level in the total program structure. Thus, a "no" answer would be right.

The authors agree with those who selected "maybe." To us a "maybe" answer implies that you understood that insufficient information was given for you to make a definitive choice. In our explanation above we've described only two alternative assumptions which could be involved in determining levels of objectives. There are many more.

The point is that we can't be sure unless we have a program structure. The heart of programming is structuring—the generation of, comparison of, and the selection of mixes leading to an integrated total program structure.

Turn to page 59

Budgeting

While we agree that Budget Method 1 touched upon some of the facets of PPBS (a goal of establishing spending limits, efficiency, etc.), it basically ignored the planning and programming aspects of establishing needs, generating alternatives, and examining trade offs. Although the various departments were given flexibility in using the resources given to them, systematic analysis of alternative objectives and activities between departments was not described in Budget Method 1. Thus we believe that this method is inadequate to the task of producing a programmed budget.

Turn to page 65

You're right. It may sound familiar, and it may employ some of the jargon associated with PPBS, but it supplied little evidence that consideration was given to the planning and programming aspects. Educational needs, alternative educational objectives, and activities were not systematically considered. Budget Method 1 failed to consider specific programmatic trade offs. Perhaps more importantly no attempt was made to relate inputs to expected educational accomplishments.

Turn to next page

Budgeting 65

BUDGET METHOD 2

This participatory method of budgeting is practiced in many of our "better" school districts. Let's listen in.

> First of all, we'll have to decide what is needed. To do so, we'll have the school personnel complete our budget forms to indicate their needs for personnel, materials, equipment, space, travel, etc. Then we'll compute the cost by adding to their estimates the costs of school board, administration, transportation, debt, and other necessary support services. Since we can't be sure that our estimates will be accurate because of inflation, unresolved contract negotiations, unforeseeable plant and equipment breakdowns, we'd also better set up a contingency reserve. After we estimate our total expenditures we'll check our revenue situation. If expenditures exceed revenues by an amount sufficient to cause a $5.00 per thousand increase in the property tax rate, we'll carefully reexamine our spending plan and reduce costs in those areas with the lowest priority in order to hold the tax rate increase at or below $5.00 per thousand. In so doing we can show the taxpayers that we've carefully developed and analyzed our budget and that we've removed all the "fat."

This method will produce a programmed budget. . . . turn to next page

This method will not produce a programmed budget. . . . turn to page 67

Sorry. Although this may be common budgeting practice for many school districts, it will not produce a programmed budget. A programmed budget requires that specific, measurable objectives be developed and then linked with their estimated costs and benefits. It also requires the development of alternative objectives and activities again linked with their estimated costs and benefits. Method 2 did not mention educational need, instructional objectives, or program objectives. It was almost devoid of alternatives. Although it mentioned that cuts were to be made on a "lowest priority basis," it did not tell us the criteria underlying the priority system.

Remember that the procedures described in Budget Method 2, represent an oversimplified version of budgeting games played by many agencies of government including school districts. But also remember that the introduction of PPBS probably will not replace these practices; it is more likely that PPBS will have to accommodate these practices. To implement PPBS will require that objectives, alternatives, cost/benefit, and trade offs be integrated with on-going budgeting systems.

Turn to page 68

Budgeting

You're right. This will produce something but certainly not a programmed budget because specific, measurable, and alternative objectives linked to their estimated costs and benefits in a systematic program structure have not been noted. Budget Method 2 is a vague but widely practiced method. It was almost devoid of alternatives. Although it mentioned that cuts were to be made on a "lowest priority basis," it did not tell us the criteria underlying the priority system.

Remember that the procedures described in Budget Method 2, represent an oversimplified version of budgeting games played by many agencies of government including school districts. But also remember that the introduction of PPBS probably will not replace these practices; it is more likely that PPBS will have to accommodate these practices. To implement PPBS will require that objectives, alternatives, cost/benefit, and trade offs be integrated with on-going budgeting systems.

Turn to next page

Budget Method 3

In this method we (1) examine and investigate educational need, (2) develop alternative, specific and measurable objectives, (3) create program structures that include alternative objective-activity sets, (4) determine the costs and benefits of each set, (5) analyze the objectives-activities mixes, (6) trade off alternative objectives-activities mixes and determine the most efficient sets of mixes, (7) select the revenue-spending plan that optimizes educational output within the constraints of anticipated revenue from all sources, (8) authorize the first year's phase of the multiyear program, (9) carry out the first year's phase, and (10) update the planned program annually, taking into account feedback from the prior year's experience.

This method will produce a programmed budget. . . . turn to page 71

This method will not produce a programmed budget. . . . turn to page 70

Budgeting

We feel strongly that budgets should go beyond their traditional functions of estimating revenues and carefully specifying the inputs that a school district can use. In addition to these functions we believe that budgets should relate the resources (inputs) of the school district to the outputs expected. Indeed, budgets should reflect the objectives and activities undertaken by the school district to improve or educate the students.

So far we have merely asserted that most school budgets do not do this. To further document this assertion, we have outlined on the next page a typical example of a budget of a school district. We have modified an actual budget somewhat to simplify its presentation; however, it still reflects an accurate picture of the actual budget from which it was drawn. Thus, assume that it is complete and realistic.

While using the budget on the next page as a reference, answer the following questions. Our answers can be found on page 73. After you have *completed* the following quiz, compare your answers to ours.

Quiz

1. How much does the K-12 reading program cost? _____
2. How much does the K-12 math program cost? _____
3. How much does the K-6 science program cost? _____
4. How much does the guidance program cost? _____
5. How much does the sports program cost? _____

Turn to page 72

Whoops! We hoped you wouldn't get to this page. In budget Method 3 we tried to outline the major events and processes we discussed from the beginning. In brief the ten steps outline the essence of PPBS. Maybe you can better demonstrate your understanding of these aspects of PPBS by applying your newly acquired knowledge to a variety of different "model" budgets developed in the next few pages.

Turn to next page

Budgeting

You are right. It may be new to some budget makers, but this is what the authors have in mind. You now seem to have at least a basic understanding of the PPB system. But will you be able to apply your knowledge? We'll give you a chance to test yourself in the next few pages. We'll briefly describe a budget of a school district and then give you an opportunity to criticize some "model" budgets.

Turn to page 69

INSTRUCTION—REGULAR DAY SCHOOL BUDGET

Supervision—Principals *(275,314)

Salaries of principals and vice principals	$186,406
Salaries for clerical personnel	61,072
Equipment, supplies, mileage	27,836

Teaching *(3,394,747)

Salaries of teachers K-12	2,953,061
Aides	50,000
Substitute teachers	67,000
Equipment	76,241
Supplies, textbooks, library books	201,295
Educational TV and conferences	30,150
Special and vocational education services	17,000

Instrumental Music *(15,700)

Instruments	9,000
Repair	6,700

Sports *(53,186)

Coaches	31,000
Equipment	6,074
Referees, mileage	16,112

Guidance *(114,813)

Counselors	82,782
Clerical	21,967
Equipment	1,644
Supplies, periodicals, mileage	8,420

Psychological Service and Health *(109,878)

Nurse, teachers, psychologist	90,683
Equipment	1,782
Other Expenses	17,413

TOTAL INSTRUCTION—regular day school $3,963,638

(*Figures in parentheses are sub-totals of the items listed below.)

Turn to next page

Budgeting 73

Here are our answers to the quiz. Do you agree with us?
1. ????????
2. ????????
3. ????????
4. $114,813
5. $53,186

We hope that you are not too surprised. There just is not enough information for the authors to answer the first three questions. If you did work out answers to the first three questions, please let us know how you did it and we will buy you a cup of coffee or some other drink of your choice if your method is valid and reproducible.

The answers to questions four and five are easy to obtain; just find the title of the program on the previous page and enter the figure listed opposite it. Eighty-two percent is not related to any meaningful label such as guidance, sports, math, science, or reading; all we know is that nearly 3.4 million dollars will be spent on teachers, their aides, supplies, and other services.

The budget that was presented on the previous page didn't just happen; a lot of judgment, information, experience, and wisdom went into the making of that budget. This is the form of the budget that is typically issued to the public (and to most of the staff and parents) for their decision either to approve or disapprove it.

For our purposes here the issue is not what the result of the budget vote was, but rather that the information given is mainly useful in controlling what kind and type of inputs will be permitted. A PPBS budget would relate the wealth of data that was used in compiling that budget in a way that would be more useful to any decision maker.

Turn to next page

Without attempting to construct a complete program budget within this text, it may be useful to recast the teaching part of the budget, the part that consumes $3,394,747 or 82 percent of the budget. A more complete program budget would include:

(1) More categories and subcategories drawn from our program structuring efforts;

(2) Well-stated, quantified, measurable objectives for each of the categories;

(3) A great deal of data about the student population and their teachers presented in the framework of the objectives in 2 above;

(4) A multiyear time frame instead of only one year.

However, we will illustrate (in Program Budgets A and B) *only* how a restatement of the budget in program terms could provide us with a clearer idea of what some of the subprograms are and how the relative cost of each could be shown. Data not shown are (1) explicit objectives for each student, class, grade level, and/or subject-matter area, (2) a multitude of information about each level of these objectives, and (3) a multiyear time frame. These data would be available as separate, supporting documentation to the budget if the budgeting process were part of a PPB system. Why? Because these are the essential ingredients of a PPB system. Without them you would not have a programmed budget. Try to imagine then that Program Budgets A and B have resulted from a PPB system and thus explicit objectives and a wealth of data about each objective are available in supporting documents.

Turn to next page

Budgeting

MODEL BUDGET A
EMPHASIZING GRADE LEVELS AND CURRICULAR AREAS AS OUTPUT OR BENEFIT CATEGORIES

Early childhood (Pre-K and K)	$250,000*
Primary grades (1-3)	250,000*
Intermediate grades (4-6)	300,000*
Reading (K-12)	400,000
Social sciences (7-12)	300,000
Math and science (7-12)	600,000
Physical and health education (7-12)	400,000
Special education (K-12)	400,000
Adult and vocational education	494,747
TOTAL INSTRUCTION	$3,394,747

*Cost of reading not included

NOTE: Often instructional programs will include prorated shares of supervision, guidance, and so on; these costs are not included above.

Turn to next page

MODEL BUDGET B
EMPHASIZING GRADE LEVELS AS OUTPUT OR BENEFIT CATEGORIES

Early Childhood Programs	(1,000,000)
Prekindergarten	$200,000
Kindergarten	100,000
Primary level	300,000
Intermediate level	400,000
Junior High Program	(1,000,000)
Attendance area "A"	500,000
Attendance area "B"	500,000
Senior High Program	(1,200,000)
Adult and Summer Program	(194,747)
TOTAL INSTRUCTION	$3,394,747

NOTE: Often instructional programs will include prorated shares of supervision, guidance, and so on; these costs are not included above.

Turn to next page

Budgeting

In Model Budget A, we have shown that a group of programs and subprograms have been clustered together into a program structure to form major program areas such as early childhood, primary grades, and so on. This program structure is built up by employing grade levels and/or curricular areas. Each of these program areas is mutually exclusive. Thus, while reading is a part of early childhood and primary grades in actual operations, in this budget it is shown as a separate entity for analytical and communicative purposes. It should be noted that this program structure is quite an arbitrary decision on our part. Other groupings may have been just as good. See for example Model Budget B. Remember that program budgets are supposed to reflect programs in which similar programs and subprograms are grouped together. However, each school district must decide for itself whether its output categories will emphasize students, curriculum, personnel, time, location, pedagogy, or something else.

Referring to Model Budget A, answer the quiz on the next page.

Turn to next page

Quiz

(Refer to Program Budget A)

1. How much does the 7–12 math and science program cost? _____
2. How much does the special education program cost? _____
3. How much does the K–12 reading program cost? _____
4. How much will the reading program cost over the next five years? _____

Turn to page 82

Tomorrow Is Built Today

As stated earlier, the three processes have the same end in view and when coupled with systems provide a rational, reasoned approach to the establishment of objectives and the selection of alternative courses of action in view of what they cost and the benefits obtained. In the course of relating inputs to outputs, it is important to look at the dimension of time. After all, the school usually interacts with the student for twelve years before he becomes an output.

Stretching one's imagination to think through everything for twelve years is extremely difficult. Conversely, thinking only a year ahead may be easier but it may not be as rewarding. Somewhere in the middle would be reasonable, for PPBS involves recognizing that today's decisions are commitments to the future. It's important in PPBS not only to worry about this year, but to look ahead. You don't just simply review and renew all of last years' activities, but decisions should be made in light of current consequences and future consequences.

The budget on page 72 is almost $4,000,000 for one year and probably would exceed $20,000,000 for a five-year period. Let's speculate for a moment about how our decisions might be influenced if we think of $20,000,000 rather than $4,000,000.

First of all, $4,000,000 quickly puts us in the mind-set of the current educational program. That is, most of it is consumed by paying for the salaries and related costs of teachers, administrators, janitors, bus drivers, cooks, instructional aides, and other personnel. Next, fixed costs for debt (the result of a building program) further narrow the range of "optional resources." Eventually we think that our "real" options are in fact limited to supplies, textbooks, field trips, and so forth.

But when we think of $20,000,000 we tend not to be so closely tied to the present program. For example, attrition of personnel in all categories over five years may produce enough savings to consider the creation of a well-designed instructional resource center which would provide us with pedagogical options that would not be otherwise available. Thus our total educational program could include a heavier reliance on individualized instruction, a much broader range of alternative curricular configurations, supplementary instruction, remedial instruction, and so on. Moreover with such a center, it would be possible to consider a more intensive use of the facility. Hence we may explore the possibility of lengthening the school day and the school year to help attain more intensive use. If the extended school day and year were found feasible, these innovations might permit the district to "save" the potential costs of new buildings, busses, and food service facilities. We could go on with other examples. We hope you understand that the time dimension together with its implications for resource availability and use is an important consideration in PPBS.

The authors in an effort to bring home the messages of the processes of planning, programming, and budgeting have discussed them, by and large, as separate processes and may have given the impression that PPBS requires that they be approached in a linear fashion. Indeed the opposite is true as can be seen by referring to our drawing above. You may feel that as you complete one process in PPBS you need not return to it. However, the component processes of PPBS are mutually dependent and all processes have the same end in view. Therefore, need will dictate the process to be engaged rather than format.

Turn to page 83

Budgeting

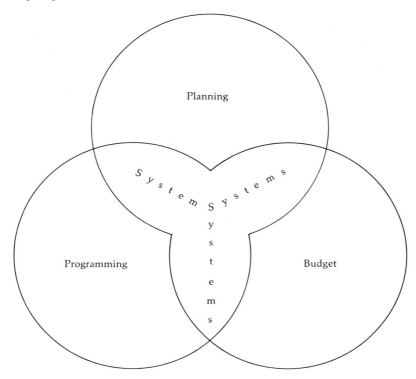

THE SYSTEM IN PPBS

ANSWERS TO QUIZ

1. $600,000
2. $400,000
3. $400,000
4. ????????

The answers to the first three questions should be easy; merely find the program in Budget A and find the dollar amount listed opposite that program. Question four cannot be answered because only one year's cost has been shown. However, PPBS requires that we analyze our programs over an adequate time horizon.

Turn to page 79

Budgeting 83

PPBS SUMMARY

PPBS is surely not the answer to every problem involving every issue in education but rather a methodology to aid decision makers in allocating scarce resources within an environment where "free" resources are disappearing. This is done through the following interdependent processes:

> PLANNING. ... The process of determining objectives (stated in quantitative terms) and specifying alternative objectives.
>
> PROGRAMMING. ... The process of optimizing the mix of inputs necessary to attain a specified objective.
>
> BUDGETING. ... The process of authorizing the use of resources to accomplish the objectives and programs developed in the planning and programming processes.

In addition, PPBS builds in a dimension over time so that the decision maker can see today's decision in terms of short- and long-term consequences. Efficiency is built into the methodology in two ways. We can seek the most value for a given cost, or try to attain a given objective for the least cost. PPBS's greatest asset is its potential for integrating planning, programming, and budgeting processes.

Turn to next page

Bring Home the Bacon

For most school districts PPBS is a supplementary type of budgeting and requires an expenditure of resources beyond what most states set as minimum budgeting and reporting standards. The budgets that are required by the majority of states are for the purposes of control of monies. They function extremely well in protecting the taxpayers of a school district against illegal expenditures. However, they are not geared to do the job that PPBS can do in helping the taxpayer get the most for his money.

Therefore, as we have stated many times over in our book, we feel that PPBS belongs in education. For those who feel the same as we do, we have provided an implementation section in this book. We hope it will serve as a guide to help with the installation of PPBS in your school. That is, BRING HOME THE BACON.

Turn to next page

Implementation Guide

YCIPPBSIYSDBSOYA

You can't implement PPBS in your school district by sitting on your alternatives . . . or, not to decide is to decide!

You've come a long way, so don't stop now by sitting on the alternatives. Move right ahead into the Implementation Flow Chart for PPBS. It presents a step-by-step guide to help you put PPBS in your school district. (And by no means is it the only way it could be done.) The Implementation Flow Chart also will give you a chance to use your newly found knowledge. Or to put it another way, you'll get to examine your alternatives rather than sit on them. But first, a few words about flow charts.

A LEGEND

Once upon a time, (at least three generations ago), there were tribes of people (programmers by name) who were responsible for the care and feeding of computers. In the very beginning, the programmers could not use words to feed the baby computers. They had to use numbers.

As the computers became bigger, they became very difficult to feed. Everything they were fed had to be in sequence. Eventually the programmers found a way to get the computers to accept wordfood, but the monsters still had to translate the words into numberfood.

To be sure that the wordfood would be translated into the proper numberfood sequence within the computer, the programmers invented the flow chart. Thus by drawing pictures (like their ancestors of old), the programmers found they could feed their computers with ease. But there's more to the legend.

When various tribes from different camps (reservations) got together for pow-wows, they discovered they had taught their computers different words. But their invention again was helpful. Flow charts were found to be useful to expedite communication from computer to computer and from reservation to reservation. Today, legend has it that these tribes of programmers have given us flow charts and computerized reservations.

ANOTHER LEGEND

⇨ A useful symbol borrowed from earlier tribes to show direction.

◯ Connector symbol showing exit to or entry from another part of the flow chart.

◇ Decision or choice symbol which allows alternative paths to be followed.

▭ A symbol representing each step or function involved in the accomplishment of the final outcome.

⟋⟍ A symbol used for showing a communication link.

Turn to the next page

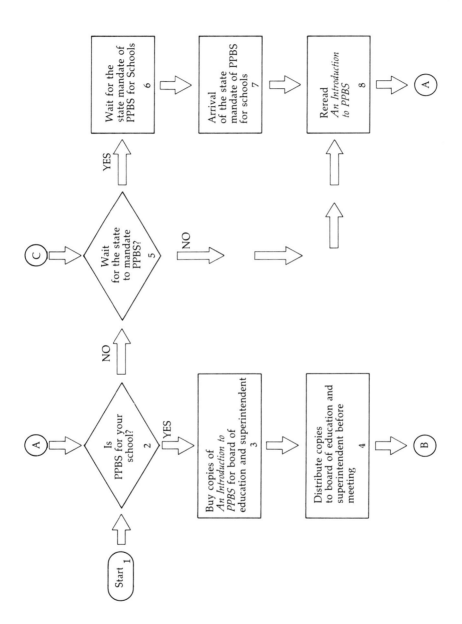

1. *Start.* . . . The most frequent question we are asked in classes, workshops, and group sessions is "how do we start PPBS." Well, our answer is that you already have started it. By reading the book to this point, you've demonstrated your interest in it. Perhaps you would like to use it at some time in the near future. Well, our focus has been on education from the onset, so the Implementation Guide has to do with the implementation of PPBS in your school system. So the real question to ask is. . . .

Turn to next page

Implementation Guide

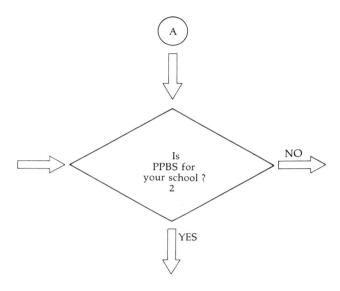

2. *Is PPBS for your school?* We hope that the answer to this question will be yes. However, before you answer the question, consider the following:

PPBS Is Not
- The answer to every problem involving every issue.
- An attempt to computerize the decision-making process.
- Just another way to save money or cut expenditures.
- A substitute for judgment, wisdom, experience, and knowledge.

PPBS Does
- Attempt to make the decision-making process explicit.
- Attempt to assure the decision maker with a choice of valid comparable alternatives.

IS PPBS FOR YOUR SCHOOL?

No. . . . turn to page 92

Yes. . . . turn to page 90

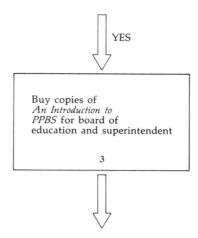

3. *Buy Copies of* An Introduction to PPBS *for the Board of Education and the Superintendent.* You may think that we are just trying to sell books. Well, although we would like to sell more books, we actually have a more important reason for putting this in the flow chart. The intent of our book was to give you the fundamental concepts of PPBS and a grasp of the jargon. In addition, we hope that we have convinced you of the merits of PPBS. We also assume that, if you have reached this step in the flow chart, we have accomplished at least part of this goal.

Using this assumption, why not let us go to work for you? Give a copy of *An Introduction to PPBS* to the board of education and the superintendent. Just the Jargon Guide alone will help to bridge the communication gap.

Turn to next page

Implementation Guide

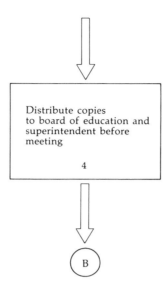

4. *Distribute Copies to Board of Education and Superintendent before Meeting.* We put this in the flow chart to stress the point that nothing much will be accomplished if those involved are not familiar with the fundamentals of PPBS. Therefore, if progress is to be made at the first meeting about PPBS, then make sure that the copies of *An Introduction to PPBS* are passed out well in advance of the meeting. Just to be on the safe side, maybe you'd better give the superintendent his copy first.

Turn to page 96

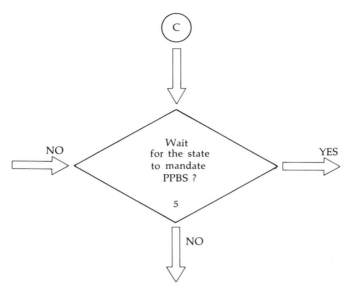

5. *Wait for the State to Mandate PPBS?* Well, this is a viable alternative to trying to interest the superintendent and board of education in PPBS right now. However, you may have some information that we don't have as to why you shouldn't have PPBS in your school. Excluding the alternatives that the board of education has already approved PPBS or that they have already investigated it and decided to wait for the state mandate, it may be just a matter of timing. So wait! However, if they are too busy with other things and have some tough decisions ahead of them, maybe you shouldn't wait for the state to mandate PPBS.

WAIT FOR THE STATE TO MANDATE PPBS?

YES. . . . turn to page 93

NO. . . . turn to page 95

Implementation Guide 93

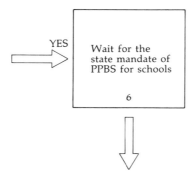

6. *Wait for the State Mandate of PPBS for Schools.* Well, YOU decided to wait based upon all the decisions you have made up to this time. You'll just have to wait for the state mandate. If you wish to pass the time boning up on PPBS, we suggest our Jargon Guide as a beginning and the references on page 7 to be followed next. No matter what you do, keep this book handy and place a bookmark in this place. *Remember: do nothing about PPBS in your school until the arrival of the state mandate of PPBS for schools**

*May we suggest an alternative. If you are truly sorry that you decided to wait for the state mandate and want to proceed to implement PPBS in your school, return to step 2 in this flow chart and select the YES answer.

Turn to next page

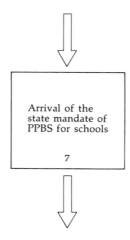

7. *Arrival of the State Mandate of PPBS for Schools.* See, we knew it would come. It may be called by another name but in essence it is still PPBS. Don't let them fool you with a lot of different acronyms and jargon. Under the umbrella of euphemisms, you probably can perceive that its putative purposes are only proxies for PPBS. Thus, read the state mandate and try to grasp the fundamentals of what they are attempting to do in their approach. If they have done a good job, you'll be able to see the dynamics of planning, programming, and budgeting at work.

Turn to next page

Implementation Guide   95

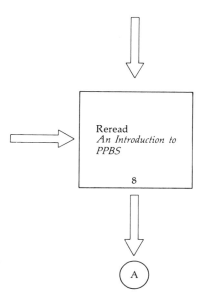

8. *Reread* An Introduction to PPBS. Since our book deals with the fundamental concepts of the modern approaches to budgeting, we think it will serve you well to review it. It is our hope that by rereading our book you will have a clearer notion of what your state mandate is trying to accomplish. In addition, it will supply you with background information that will be helpful in implementing the state mandate. When reading the rest of the Implementation Guide, you can disregard all NO answers since PPBS has been mandated by the state and you no longer have a choice about developing a PPB system for your school district. So return to step 2 and select the YES option.

Turn to page 89

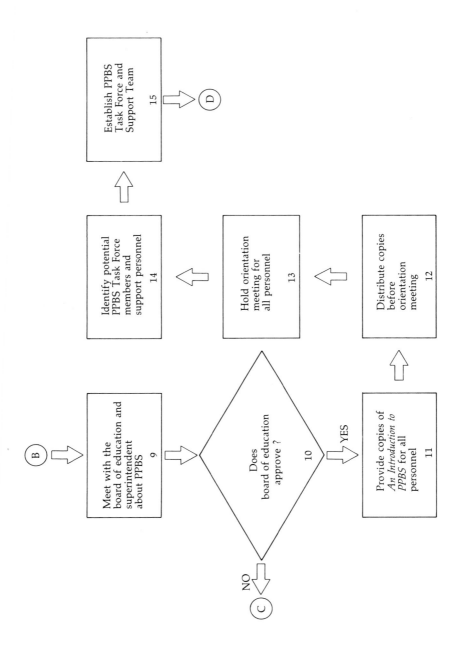

Implementation Guide 97

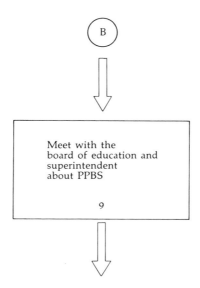

9. *Meet with the Board of Education and the Superintendent about PPBS.*
The main idea here is to have the meeting about PPBS and nothing else. We think that PPBS deserves at least that much.

Although the ingredients of PPBS are largely not new, their arrangement and use are. Thus, there will be new considerations for the people at this meeting. After all, they may be still thinking in a traditional way. You might get them out of their rut by asking them a few questions such as:

> Specifically, what are the accomplishments of existing programs?
> What are the future costs of these existing programs?
> Are today's decisions made with knowledge of their long-term consequences?

GO GET THEM TIGER. . . . Tell them about PPBS and what it can do for a forward looking board of education and superintendent.

Turn to next page

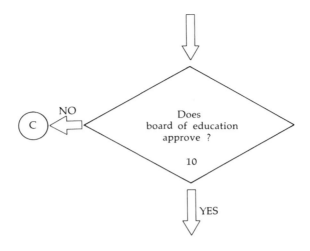

10. *Does the Board of Education Approve?*

While PPBS attempts:
- –To take account of all costs inherent in decisions,
- –To relate the wealth of data available on almost any subject or issue in a way that is useful to decision makers, and
- –To build in a dimension over time that tries to see today's decisions in terms of their longer term consequences,

It does require the school system to have:
- –A specialized staff carrying out continuing in-depth analysis of the school's objectives and its programs to meet them,
- –A multiyear planning and programming process,
- –A budgeting process which can take broad program decisions and translate them into the context of a detailed budget submission, and most importantly,
- –STRONG ENDORSEMENT BY THE BOARD AND SUPERINTENDENT.

DOES THE BOARD OF EDUCATION APPROVE?

YES. . . . turn to page 99

NO. . . . turn to page 92

Implementation Guide

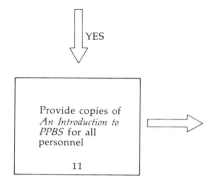

11. Provide *Copies of* An Introduction to PPBS *for All Personnel.* In view of the fact that you are on the way to implementing PPBS in the school system, we think that we can still perform a service for you by introducing the fundamental concepts of PPBS to all personnel. We believe this will save you time. In fact, we are convinced of it.

The authors have a mutual friend who does consulting with some rather large firms. Invariably at the close of the first group meeting, he quickly points out the basic problem is that all members of the group have failed to grasp the fundamentals. He generally suggests that he run a workshop before they proceed any further. (For an additional fee, of course.)

We believe that everyone should go at his own pace, so. . . . PROVIDE COPIES OF *An Introduction to PPBS* FOR ALL PERSONNEL.

Turn to next page

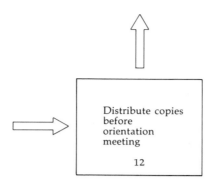

12. *Distribute Copies before Orientation Meeting.* At the risk of repeating ourselves, get the copies of the book out early to the central office staff, building principals, department chairman, and selected teachers. Give them plenty of time to read *An Introduction to PPBS* in advance of the orientation meeting.

Note: If the state mandate is in by now, send it along too.

Turn to next page

Implementation Guide 101

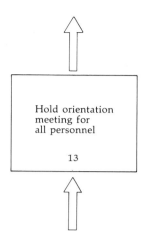

13. *Hold Orientation Meeting for All Personnel.* After a brief general meeting (it will be brief because they will have been introduced to PPBS), we suggest that the participants be grouped according to what they will contribute to the installation of PPBS in your school system. Each group should attend an activity-centered workshop to provide them with the necessary knowledge and skill to implement PPBS.

The workshops can be on any phase of PPBS and contain such activities as writing performance objectives, structuring activities, cost-benefit analyses of mixes of inputs, or developing a program budget. Whatever is decided for content, we suggest that the workshop be centered around a specific activity so that initial experience can be gained by all personnel.

Turn to next page

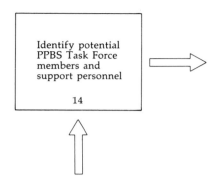

14. *Identify Potential PPBS Task Force Members and Support Personnel.* Although the power or authority to appoint staff to the PPBS Task Force remains with the board of education and the superintendent, it would be helpful to identify potential task force members and support personnel. Essentially, this is an instance of need identification that should be familiar to you from our discussion of systems. In this case, instead of asking what are our needs, we ask whom do we need.

Turn to next page

Implementation Guide

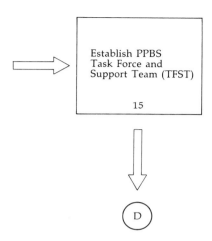

15. *Establish PPBS Task Force and Support Team.* The superintendent appoints personnel to the PPBS Task Force and Support Team (TFST). He should also appoint a chairman unless he wishes to assume the chairmanship himself. In addition, the superintendent should formally charge the Task Force and Support Team (TFST) with its responsibilities.

Turn to next page

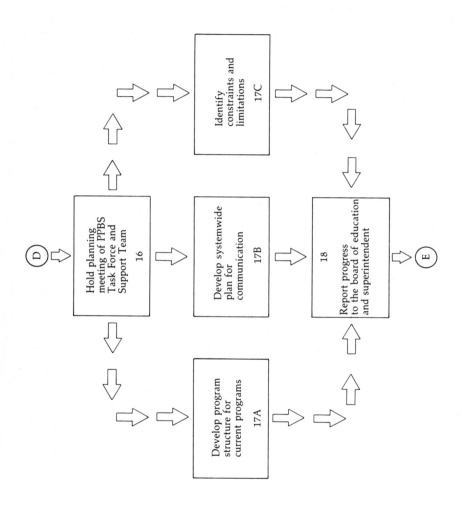

Implementation Guide 105

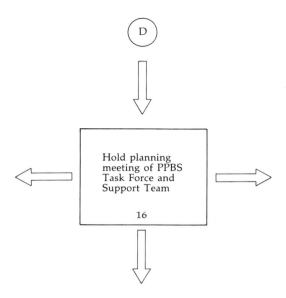

16. *Hold Planning Meeting of PPBS Task Force and Support Team (TFST).* The initial task at the planning meeting should be the development of a plan of operations, a time schedule, and the assignment of individual or group tasks. The authors suggest that a PERT Network might be developed to provide for effective management control of the team efforts.

Turn to next page

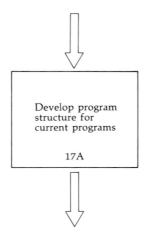

17A. *Develop Program Structure for Current Programs.* * The PPBS Task Force and Support Team analyzes the current array of activities and procedures in the school system to tentatively describe and understand them. Then, through the process of structuring, it develops a hierarchy of current programs. Thus current programs will be structured by levels and will facilitate the identification of a target program or programs. A target program is a potential program for further development and application of PPBS.

*This activity or function takes place concurrently with activities 17B and 17C.

Turn to next page

Implementation Guide 107

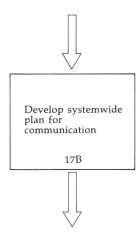

17B. *Develop Systemwide Plan for Communication.** The PPBS Task Force and Support Team should evaluate and analyze the current district communication procedures. A comprehensive plan should be developed for effective communication with the entire community (taxpayers, students, and district personnel). Information on PPBS and the progress of implementation should be provided to the community throughout its installation.

*This activity takes place concurrently with activities 17A and 17C.

Turn to next page

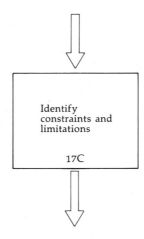

17C. *Identify Constraints and Limitations.** While the board of education has agreed to proceed with the installation of PPBS in your school system, it may not be totally aware of factors that may inhibit its implementation. Any economic, political, social, or other constraint should be identified and explained. Remember, early identification of potential or actual problems will permit you to explore them and develop strategies for a solution; it may also save you time and money later on in the implementation of PPBS.

*This activity takes place concurrently with activities 17A and 17B.

Turn to next page

Implementation Guide 109

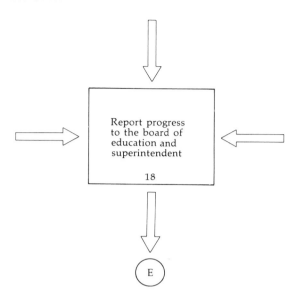

18. *Report Progress to the Board of Education and Superintendent.* When the PPBS Task Force and Support Team have (1) developed a program structure for current programs, (2) developed a systemwide plan for communications, and (3) identified constraints and limitations, they should present a progress report to the board of education and superintendent. The report should include the product of activities 17A, 17B, and 17C. It should not only outline next steps to take in implementing PPBS but also supply recommendations of possible target programs.

Turn to next page

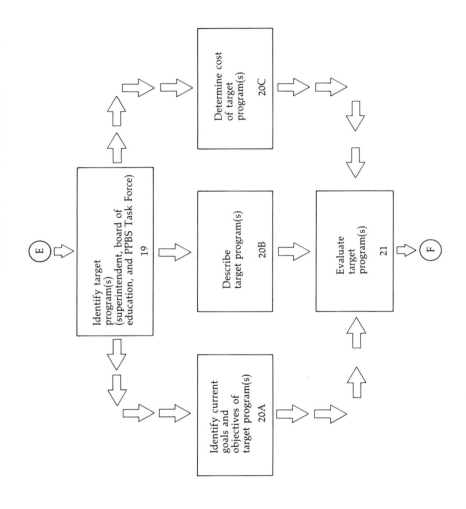

Implementation Guide 111

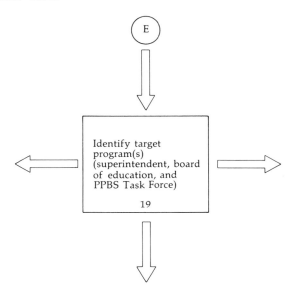

19. *Identify Target Program(s).* The identification and selection of target program(s) should be done at the progress report meeting or at a subsequent meeting. As another alternative, the board may simply react to the progress report, provide its inputs and recommendations, and direct the superintendent to select the initial target program(s).

We suggest you carefully file the list of target programs that were not selected. It may be of use later on in the flow chart.

Turn to next page

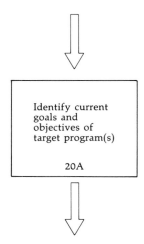

20A. *Identify Current Goals and Objectives of Target Program(s).** This begins the description of what you have now. The personnel currently involved in a target program must have a voice in determining the goals of the program and the specific objectives they are currently trying to achieve. It is not necessary that the initial objectives be stated in qualitative and quantitative terms. It is more important that they be delineated and communicated by the personnel responsible for the target program. We suggest that they use resources such as curriculum guides and courses of study in order to be as comprehensive as possible. The group should also mutually exchange individual goals and objectives for the purposes of obtaining clarity and concensus. The final product of this activity should be a clear statement of goals and specific, explicit, and measurable objectives related to each goal.

*This activity takes place concurrently with activities 20B and 20C.

Turn to next page

Implementation Guide 113

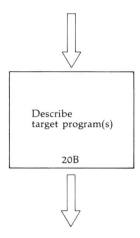

20B. *Describe Target Program(s).** Well, you may have guessed it by now. What comes after planning (goals and objectives)? Programming, of course! In this case, we will use structuring to help us along. That's right, it is who gets what by whom, and also considers when, where, and how it will take place.

It may be a good idea to refer the personnel involved to our structuring section and use it as a guide. By moving directly from the objectives, they should be able to fill in the who and what blocks immediately. And the objectives may help to suggest the descriptors needed to fill in the rest of the blocks. Structuring of current programs should be easier than structuring alternative ones because it merely requires recording what is, not what could be.

*This activity takes place with activities 20A and 20C.

Turn to next page

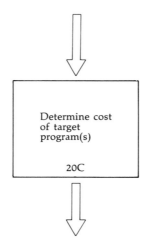

20C. *Determine the Cost of Target Program(s).* * As each program is structured, the group should start to build toward a program budget. Again, we would like to suggest that our programming and budgeting procedures be used as a guide. In this step, the group will develop a program budget structure and convert data from the current operating budget to a program budget reflecting the uses of "real" and monetary resources. The authors realize that this is not always an easy task and that sometimes assumptions must be made in order to determine costs. If such assumptions are made, they should be explicitly stated in the program budget or in its supporting documents.

*This activity takes place concurrently with activities 20A and 20B.

Turn to next page

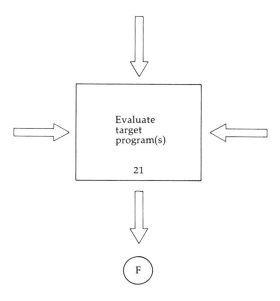

21. *Evaluate Target Program(s).* This is just another way of asking if you have been getting the punch for the penny that you thought you were. Is the program achieving the specified objectives and goals that were stated in activity 20A? Are the goals and objectives realistic for your population(s)?

While each school system may desire to establish its own evaluation procedures, we recommend that such procedures include safeguards to insure coordination among the groups involved. Early agreement on the meaning of evaluative information and procedures may minimize potentially, dysfunctional consequences later on.

Turn to next page

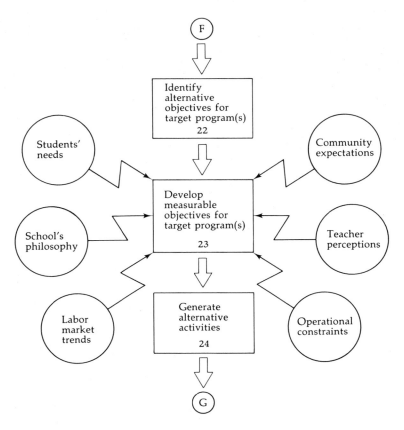

Implementation Guide

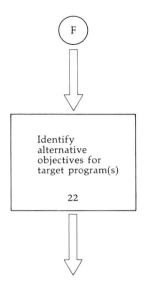

22. *Identify Alternative Objectives for Target Program(s).* Well here we have returned to planning. A good basis for beginning this procedure may be the results of your evaluation of target program(s). There is a high probability that alternative objectives may be needed. Generally, from our experience, we have found that this is almost always the case. To the extent the target program(s) is not meeting the needs of the district, we suggest that efforts be made to broaden and intensify your search for alternative objectives. To do so you may want to extend your investigation of need to include the gathering of information from the persons, groups, and other sources identified in step 23 on the next page.

Turn to next page

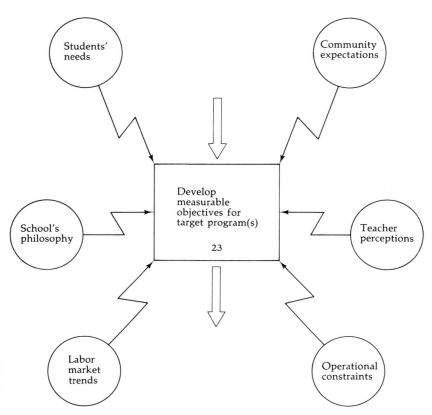

23. *Develop Measurable Objectives for Target Programs.* At this point in the implementation process the target program personnel in conjunction with the Task Force and Support Team should be prepared to go beyond the mere description, explanation, and costing out of the current target program(s). Brainstorming, community surveys, reviews of programs in other districts, and so on, should be useful in obtaining inputs to develop new and realistic objectives in qualitative and quantitative terms. The planning team should also be aware of the time dimension. Hence the objectives should reflect reasonably obtainable outcomes for the immediate and the longer range future.

To indicate that the development of alternative objectives should not be done in isolation from the real world, we've indicated (in the flowchart) that such factors as student needs, school's philosophy, labor market trends, community expectations, teacher perceptions, and operational constraints be carefully considered.

Turn to next page

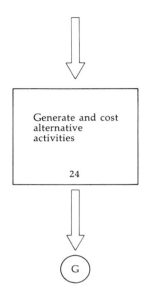

24. *Generate and Cost Alternative Activities.* After the objectives have been agreed upon, the group should move from the planning process to the ... programming process. More specifically, you may recall our discussion of program structures; we suggested in phase one, an approach which should help you begin this event. By using such an approach those involved in your target program(s) should be able to structure several alternative mixes for a given objective.

As indicated in the text, for each mix there should be an estimate of cost. Usually, the skills of specialists (accountants, economists, etc.) are needed in order to determine the cost of each mix.

While some advocates of PPBS believe that virtually all possible alternatives should be examined, this is far too inclusive for the authors. We feel that the knowledge, experience, wisdom, and judgment of district personnel together with the advice and feedback from outside consultants will help reduce the number of viable alternatives to a reasonable amount.

Turn to next page

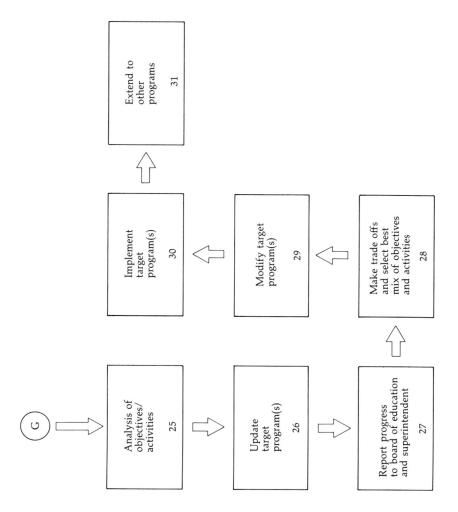

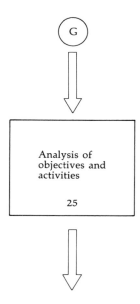

25. *Analysis of Objectives and Activities.* To help you with this event we refer you to our discussion of program structure, phase two. In order that objectives can be compared and the most efficient mix selected, it is essential to develop a program structure. Thus, the interrelatedness of objectives and mixes at comparable levels as well as at different levels within the school district's target program(s) can be examined.

By developing a program structure, it can serve to help you reduce the number of alternatives to a reasonable range of viable objectives and mixes. The decision-making process of making trade offs is thereby facilitated.

Turn to next page

Implementation Guide

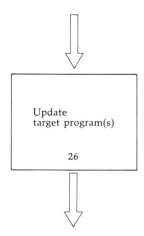

26. *Update the Target Program(s).* This is putting it all together. Now is the time for all PPBSers to update data banks. Once again our program structure should be helpful in organizing the data so that it is easily accessible.

While you are updating the data, include documentation of decisions, assumptions, and criterion rules that were employed. Include the advantages of the program, levels of programs, and mixes of elements. Keep records of what you didn't select as mixes, or objectives. Document cost allocations for all objectives and mixes (those selected and not selected).

The name of the game is to be all inclusive. Get it down and organized for the long-term haul. (You won't have to do it again.) When you finish the data banks, try them out as a basis for selecting the information to be included in your report to the board and the superintendent.

Turn to next page

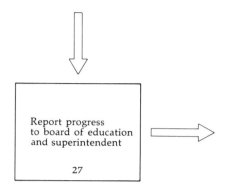

27. Report Progress to the Board of Education and the Superintendent. This activity is a major checkpoint for the results of the efforts to date. At this time the new goals, objectives, and selected mixes are presented to the board of education and superintendent along with the documentation of how the selection was made. Carefully costed objectives and resource mixes should be presented in a multiyear time frame. Goals, objectives, and mixes that were not recommended should be included in the report. (Remember, being prepared in PPBS is always having alternatives.)

From our experience, a job well done here would include recommendations concerning the extension of PPBS to other programs. If you remember our reminder in step 19 of this flow chart (identification of other target programs) you may now be able to use it. Update and refine that list and give your recommendations for the extension of PPBS to other programs in the school district. Out of this step should come suggestions and recommendations for the board and superintendent. Be sure to carefully consider these inputs in the next step.

Turn to next page

Implementation Guide

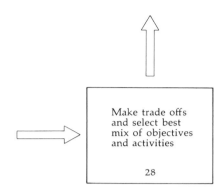

28. *Make Trade Offs and Select Best Mix of Objectives and Activities.** This is the decision-making process that is engaged in to get the most bang for the buck. Here objectives, alternative objectives, mixes, and alternative mixes are traded off to find the optimal mix of programs and subprograms. You try to get the most benefits for the program within the constraints (economic-social-political) that were originally identified in step 23 in this flow chart.

To do this the advantages and disadvantages of sets of objectives and resource configurations are weighed against other such sets. Objectives and mixes are traded off until the best configuration of these within the program structure is reached.

Trade offs involve our concept of efficiency. The final configuration of the program may not produce the highest outputs (bang); and it may not minimize cost (buck). It should, however, produce the highest output within the range of available resources and other constraints, that is, bang for the buck.

*We realize that the final decision rests with the board and thus this step could have been displayed in a decision diamond. However, we have chosen to view this step as a complex of interactions among many persons and groups ultimately leading to consensus.

Turn to next page

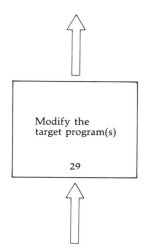

29. *Modify the Target Program(s).* Here we refine our target program(s) to reflect the inputs from the board and the results of our trade offs. This is the last formal revision of the target program(s) before it is implemented. A program memorandum should be written which includes the new mixes of goals, objectives, activities, and resources covering a multi-year time frame. The program memorandum should include plans for implementation and specifications for evaluation of the target program.

The product of this step is the program memorandum which includes a program description, an implementation plan, and the specification of how and when evaluation will take place.

Turn to next page

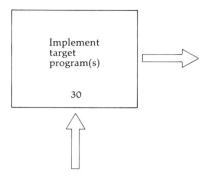

30. *Implement Target Program(s).* The target program(s) can now be implemented. The plans for implementation should be adhered to as closely as possible with evaluation taking place as specified. Barring acts of God, war, and/or civil insurrection (including student and taxpayer revolts) this activity should run smoothly and be entered into with a high degree of confidence. After all, this is what we set out to do, implement PPBS in our school system.

Turn to next page

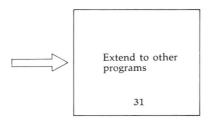

31. *Extend to Other Programs.* We assume that the board of education and the superintendent favorably considered your recommendations for the extension of PPBS to other programs in step 27 and have given you positive feedback. So you can now proceed to formulate plans for the extension of PPBS to new target program(s). Since you will be faced with development problems similar to those in the original target program(s) we suggest that you begin again at step 11. This time it should be easier because you should have access to the experience and knowledge of homegrown experts in PPBS. . . . Carry On!

Turn to next page

Jargon Guide

We realize that one feels two basic drives when new knowledge is obtained.

One is to communicate that knowledge in a useful way. Sometimes this can be difficult when trying to communicate in the presence of an expert. He generally has a set of phrases or terms that he uses with special meaning assigned to them—a second language or language within a language if you please. Generally this causes frustration to both parties, novice and expert alike, and communication breaks down. Hence, a guide to the jargon is needed to help bridge the communication gap.

The other drive is to learn more about PPBS to make yourself feel secure with the newly acquired knowledge. In our effort to help you, we have gone through a great deal of literature and devised the Jargon Guide to satisfy both drives.

As a bonus feature for the Jargon Guide, we have included some PPBS graffiti that lends its own meaning. Here we go!

Turn to next page

Alternatives. Generally it means other objectives, methods, or programs besides those already in use or decided upon. It suggests a comparison of two or more objectives, methods or programs or sets thereof. Such comparisons become the basis for trade offs.

> It's better to choose and lose than to have no chance at choice.

Bang for the Buck, et al. See Efficiency, below.

Benefit. A benefit is a planned objective that is attained. In practice it's often an objective you muddle into along the way—unplanned but the alternative that was in fact followed. Benefits can also be viewed as the outputs of a program whether negative or positive. Negative benefits are costs saved and positive benefits are the net value added to the student(s) through the educational process. (See Output, below.)

> Good and bad, there's two ways to it:
> it's what'cha get, when you do it.

Benefit/Cost Ratio. To use a benefit/cost ratio, the benefits and costs must be converted into monetary (dollar) terms. By dividing the monetary benefits by the monetary costs a benefit/cost index or ratio may be derived. When applied to alternative sets of objective-activity mixes, the one with the highest ratio indicates the mix with the highest benefit. (See Cost and Cost Effectiveness Ratio, below.)

> Bang over buck
> eliminates pot luck.

Budgeting. It is the process of legally authorizing the use of resources or inputs by a public or private organization. In PPBS, budgeting also tries to give financial expression to the objectives and programs developed in the planning and programming processes.

> Planning states purpose;
> Program the mix.
> Without Budgeting
> The System is sick

Cost. Simply stated it constitutes the specific resources (inputs) required to achieve a given output or objective. It can be quantitatively measured in two ways. First, one can count up the particular number of teachers, square feet, students, and so on, and arrive at a quantified total for each. More often, however, cost is measured in monetary terms (dollars) because a monetary index (dollars) permits one to intermix the cost of teachers with the cost of space with the cost of students, and so on.

> Assess your assets, count your costs,
> Monetize your mixes, COST! COST! COST!

Jargon Guide 131

Cost/Effectiveness Ratio. To use a cost/effectiveness ratio, inputs and outputs should be converted into quantified, measurable terms. Costs should be converted into monetary terms (dollars); outputs should be converted into numerical or percentage terms. By dividing the costs by a quantified measure of the outputs, a cost/effectiveness ratio may be derived. In this case the lowest ratio indicates the mix with the highest output. Because of the difficulty of converting all educational outputs into monetary terms, cost/effectiveness ratios can be applied much more frequently than can benefit/cost ratios. However, when different measures (nonmonetary) of outputs are employed, it is more difficult to make comparable analyses between and among alternative objectives and programs. (See Benefit/Cost Ratio, above.)

> When you've got options, you must make selections
> buck over bang helps with inspection.

Depreciation. Depreciation is the term we use to describe a reduction in the value of material resources such as buildings and equipment. Usually these resources are called assets. As assets age, they tend to deteriorate or become obsolete with improvements in technology. Things just do not last forever. In accounting, it is a systematic approach to adjusting for the loss in value of assets over their estimated useful life.

> While it's new it's nice; when it's old it's not;
> its worth declines in time; it ain't what it's not.

Effectiveness. In simple terms, it is output resulting from an activity or program. Ideally, it is a quantitative measure of an output which can be used to evaluate progress toward a goal or objective. (See Output, below.)

> Do your thing, if it's your bag,
> as long as it has a measurable tag.

Efficiency. It is that mix or set of mixes (resources such as people, materials, equipment, buildings, etc.) which results in maximum outputs, benefits, or effectiveness. Conversely, it can represent the minimum cost (lowest use of resources) at which a specific output or program objective can be achieved. Thus whether one seeks either maximum outputs or minimum cost he strives to select that mix of resources (inputs) that produces "more bang for the buck."

> The most for the least is a name of this beast.

Evaluation. Since the systems discipline is an integral part of putting it all together in PPBS, it is necessary that evaluation be ubiquitous and omnifarious. This helps insure that P, P, and B will be a dynamic, efficient process. Whether applied to past, present, or prospective relationships or sets thereof, evaluation is the process employed to audit or check out our progress toward our outputs, or goals and objectives. Decoded, evaluation simply means checking continuously, constantly, and carefully; when you've finished check again.

> When you get to the end of your rope, look again!
> It's also the beginning.

Goals. Broad and generalized statements for an organization or agency that provide a description of future states of affairs. Thus goals provide a description of the general direction(s) in which an agency or group might focus its efforts.

> Goals are wants that everyone adores.
> By stating them in objectives you're sure to score.

Interest. It is the cost one pays for the use of money. Interest rates vary in relationship to the degree of risk involved on the part of the borrower. Many factors are considered in determining risk. A "good" risk may be able to borrow at the rate of 3 percent; a "poor" risk may have to pay a rate of 15 percent (or even more).

> Tho' money may be evil,
> even it has its price.
> In discounting interest
> think about it twice.

Inputs. Resources utilized to achieve selected or defined outputs. Included are money, manpower, land, material, equipment, and other resources.

> The grist for the mill,
> the mash for the still;
> whatever goes in
> should fill the bill.

Model. A representation of a set of relationships thought to define a situation(s) under study. The "rigor" of models vary from those describing certitude of the relationships (usually mathematical) to those only verbally describing said relationships. It is generally thought that models permit one to manipulate one or more of its elements to determine or estimate their effects on each other in different situations. (See Program Structure, below.)

> A way to put your whole world in your hands.
> Think about it!

Objectives. These are specific and explicit statements of goals. Objectives are measurable and specify the quality and quantity of output(s) within time limits.

> The streets of goals are paved with objectives.

Operations Research (OR). A term frequently used but often with different meanings. Generally, it refers to the many techniques available to help solve practical or operational problems. Among such techniques are guessing techniques such as Delphi and Monte Carlo, information theory, linear programming, probability theory, Program Evaluation and Review Techniques (PERT), and other so-called Critical Path Methods (CPM) of project or program management.

> OR is thought to be very scientific.
> Always ask the user to be very specific.

Outputs. These are the expected or actual result(s) of an activity or set of activities. In planning they are expected or anticipated; in auditing, they are actual results. They are equivalent to benefits. (See Benefits, above.)

> What'cha get or what'cha got
> from what'cha did with what'cha bo't.

Planning. It is the identification and selection of goals and the specification of objectives and alternatives thereto. It starts with the identification of need—the gap between what is and what should be. The identification of needs leads to the establishment of goals and objectives which (if met) should reduce or eliminate the "need gap."

> Begetting goals for new found needs,
> operationalizing objectives does the deed.

Program. It is the mix or mixes of resources (also referred to as activities or methods) that an agency employs to achieve its goals and objectives.

> The product of postulating pieces for peak
> payoff from the peso.

Programming. It is the process of examining alternative mixes of resources (also referred to sets of activities) that can be used to achieve goals and objectives, and then selecting the mix (or mixes) that is most efficient.

> A priori analyses of alternatives for attainment.

Program Structure. A program structure is a model in which the relevant elements are carefully described and interrelated to each other. It normally includes hierarchial levels in which each lower level is a more detailed and complete description of the level immediately above. (See Model, above.)

> Levels and elements need nomenclature;
> get yourself a program structure.

Systems Analysis. It is the application of the system's discipline to solve problems. It requires the determination of needs, goals, and objectives, and the development and analysis of alternatives including continuous evaluation and feedback.

> Educational problems can cause paralysis.
> We prescribe systems analysis.

Trade Offs. Trading off is the process of examining alternative sets of objectives and/or mixes of resources (also referred to as activities or methods) and then determining which set and/or mix represents the most efficient way to meet the overall goals and objectives of a program or an agency. Thus, the sets or mixes that were *not* selected are said to "be traded off" for the one or ones that were selected.

> Weighing the ways and means to obtain the
> odds for the ends.

Evaluation: Self-Test

On the pages that follow, you will find a short self-test by which you can check your grasp of the critical concepts that have been presented in this book. The self-test is so constructed that you will have a choice of alternatives (what else?) for each question.

By making the assumption that a particular question will not stimulate recall of knowledge, we have included the alternative of Huh?. If you choose the alternative of Huh?, turn to page 138 and references will be given for that particular question. Read the references and come back and choose another alternative.

Proceed through the twenty questions on PPBS until you have selected either the Yes or No alternative for each question.

Complete the following, before taking the self-test:

Since my goal was to learn about the critical concepts of PPBS, my objective is to get _____ (fill in number) of questions right while only utilizing _____ (fill in number) Huh?'s.

Turn to next page

Twenty Questions on PPBS

	Yes	No	Huh?

1. PPBS is nothing more than a methodology for improving decisions that have to do with the allocation of resources to attain satisfaction of our wants
2. Efficiency in PPBS is simply the saving of money?
3. In PPBS, efficiency and goals are related?
4. The systems approach is used in PPBS?
5. Need is the discrepency between the present state or condition and what is intended?
6. Goals are determined from objectives?
7. PPBS eliminates alternatives?
8. Planning is generally brought about by need?
9. Planning is merely the setting down of goals and determining objectives?
10. In programming, structuring helps generate possible mixes of inputs?
11. Structuring facilitates the development of program structure?
12. Programming is dependent upon planning?
13. In programming, the decision maker trades off mixes—never objectives?
14. Seventh-grade spelling is always at a higher level in a program structure than seventh-grade spelling at the J.F.K. School?
15. The best way to structure a program is by "What's"?
16. Budgeting is related to programming, not to planning?
17. PPBS is mainly a way to save money?
18. Planning, programming and budgeting is a linear and static process. Once one phase is completed, you need never return to it?

Evaluation: Self-Test

19. PPBS is best applied to yearly budgets with its short-term consequences?
20. Evaluation in PPBS is done only at the end of the year when final tests are given to the students?

Turn to page 138 for correct answers

Correct Answers for Self-Test

1. No
2. No
3. Yes
4. Yes
5. Yes
6. No
7. No
8. Yes
9. No
10. Yes

11. Yes
12. Yes
13. No
14. No
15. No
16. No
17. No
18. No
19. No
20. No

If you didn't meet your objective (i.e., goal), write an alternative objective for the objective on page 135 and then proceed to page 1.

Evaluation: Self-Test

PAGE REFERENCES FOR HUH? ANSWERS

1. pp. 7-9
2. pp. 7-9, 89, 131
3. pp. 10, 11, 45, 125
4. pp. 11-14, 30, 82, 131
5. pp. 15-17, 29, 102
6. pp. 18-20, 112, 132, 133
7. pp. 21-27, 117-120, 130
8. pp. 30, 88-128
9. pp. 30, 45, 83, 133
10. pp. 50-59, 109, 113, 120, 122
11. pp. 55-59, 120, 122
12. pp. 52-60, 82, 83
13. pp. 57, 125, 134
14. pp. 55-59
15. pp. 55-59
16. pp. 60-61, 82, 130
17. pp. 60-71, 80, 89
18. pp. 82, 85-128
19. pp. 74, 78-81, 98
20. pp. 28-30, 122, 131

Read referenced pages and return to the Quiz on pages 136-137